The Operational Art

The Operational Art

Developments in the Theories of War

Edited by
B.J.C. McKercher
and Michael A. Hennessy

PRAEGER

Westport, Connecticut
London

Library of Congress Cataloging-in-Publication Data

The operational art : developments in the theories of war / edited by
B.J.C. McKercher and Michael A. Hennessy.
p. cm.
Includes bibliographical references and index.
ISBN 0–275–95305–X (alk. paper)
1. Operational art (Military science) I. McKercher, B.J.C.
II. Hennessy, Michael A.
U163.064 1996
355.4—dc20 95–52704

British Library Cataloguing in Publication Data is available.

Library of Congress Catalog Card Number: 95–52704
ISBN: 0–275–95305–X

First published in 1996

Praeger Publishers, 88 Post Road West, Westport, CT 06881
An imprint of Greenwood Publishing Group, Inc.

Printed in the United States of America

The paper used in this book complies with the Permanent Paper Standard issued by the National Information Standards Organization (Z39.48–1984).

10 9 8 7 6 5 4 3 2 1

Copyright Acknowledgments

The editors and publisher gratefully acknowledge permission to use excerpts from the following:

William E. DePuy. "FM 100-5 Revisited" in *Army* 30, no. 11, 12–17.

Donn A. Starry. Interview by John L. Romjue, TRADOC History Office, 19 March 1993.

Contents

1

Introduction

M. A. Hennessy and B.J.C. McKercher

"The art of war," said Napoleon, "is a simple art; everything is in the performance."[1] The conduct of war has often proven more difficult than Napoleon allowed. Since Napoleon, war in practice has undergone fundamental transitions in scale and scope associated with the fielding of mass national armies and the mass production of war matériel.[2] Finding the formula for success on the battlefield in the age of industrial war has preoccupied military planners ever since. Despite retention of the accoutrements, the development of mass armies obviated the ability of the heroic commander on horseback to manage the battlefield. The scale of modern war demanded new levels of command organization and planning. Throughout the post-1945 Cold War, the adoption of nuclear weapons as the pillar of Western defense equally obviated the study of such issues as the problems of war in the age of mass mobilization and the articulation of armies in war. Publication of the U.S. Army's new field manual, FM 100-5 *Operations,* in July 1976, fanned an intellectual renaissance in military thought that eventually resulted in the widespread adoption of the term "operational art." The Twenty-First Annual Military History Symposium of the Royal Military College of Canada, held in March 1995, marked an effort to assess the legacy of the new operations manual on mainstream military thought by examining its historical and transnational antecedents. In particular, the conference reexamined the origins of the concept and the diverse interpretations it has received.

The term "operational level" refers to an intermediate phenomenon existing between discrete tactics and wider strategy. The term "operational *art*" generally refers to the practice of generals—or their staff colonels—for achieving operational success. According to Soviet military theory, operational art was onc of three components of the military art, the connecting link between strategy and tactics. The operational art was called on to work out the theory and practice of preparing and conducting operations. In the Soviet view, the operational art remained distinct from doctrine and strategy. The former concerned the nature of future war, while the latter most importantly represented high policy and purpose.[3] The term "operational art" coined by Soviet military writer General-Major A. A. Svechin in the 1920s applied to the imaginative leadership skills required to campaign successfully on the greatly expanded battlefield of the industrial age.[4] The new demands had not been lost on others, and the concept of the operational level clearly predates the phrase.

In the period between the late 1850s and 1914, the concept of operational art was developed by the Great German General Staff. Under Field Marshal Helmuth von Moltke the first modern general staff system developed.[5] One purpose of the staff was to plan the maneuver to the battle area of divisions and corps formations to the point of contact with the enemy. Other elements of the staff considered means of fighting the battle and preparing for subsequent ones. It has been said that the transformations of warfare made possible by industrialization and mass conscription required a modern, German-style staff structure.[6] Whatever those demands, the German staff system became the model for others. The Prussian wars of expansion in the 1860s and Franco-Prussian War demonstrated the success of these planning efforts. Success validated the Prussian model of planning and study.

The staff provided the forum for a disciplined, thorough study of the nature of war and, in particular, the requirements of modern campaigning. With the creation of the early German General Staff there was recognition of the need to gather an accurate account of operational experience by means of a dedicated army historical section. Digesting and analyzing for lessons learned proved more problematic. Despite well-developed theory, the test of battle demonstrated practical incapacities. Even so, after the Franco-Prussian War, the Prussian model undeniably influenced many continental powers. Most sought to emulate the means of preparing for war innate to the general staff system and developed the historical reporting and intelligence apparatus to help achieve campaign success. Neither, of course, compelled success.

For didactic purposes, the German General Staff after Moltke employed the study of Hannibal's remarkable success at Cannae—a classic battle of envelopment. But in contrast to Cannae the battlefield through the late nineteenth century witnessed a tremendous increase in the lethality of modern weapons and the expansion of the battle area to far greater depth and breadth—trends which have continued—which immediately complicated the commander's effort to shape a winning campaign. Field Marshal Alfred von Schlieffen, the champion of the Cannae studies, was also the master craftsman behind the disastrous German offensive against France in 1914. The failure of the German offensives in 1914 and the subsequent stalemate proved that simply preparing efficient plans for mobilization and deployment were not sufficient in themselves to ensure success on the battlefield. Such setbacks invalidated neither the staff system nor the search for a successful operational art.

The following chapters concern how armies have sought to define and prepare for operational success. John English begins the collection with a broad overview. The discrete experiences of Wilhelmine Germany are then reviewed by Bradley J. Meyer. Jacob Kipp addresses the adoption of the term by the Red Army, linking its origins to both German theoretical work, particularly that of Carl von Clausewitz, and the experiences of the Tsarist and Red Armies. William McAndrew surveys the Canadian experience, in which a small army caught between British doctrine and American technology strove toward an operational art during the Second World War. Sabine Marie Decup reviews the French operational experience since 1945. The colonial retreat brought many battles, and indeed open warfare, but the French armed forces operated essentially *sans l'art opérational* and, as other armies, neglected the subject once nuclear weapons appeared to revolutionize warfare. David M. Glantz addresses the later Soviet experience, particularly the rediscovery of the operational art and classical strategies articulated before Stalin purged the officer corps and before the nuclear revolution appeared to obviate the subject of planning for successful battlefield campaigns. Richard M. Swain addresses the American experience—born of the bitter lessons of Vietnam—which drew on Soviet theories of war fighting to bring about a renaissance in military tactical/operational thinking. Stéphane Lefebvre and his colleagues move beyond that historical account to examine the implications for future force structures and requirements born of this revolution in military affairs. Charles F. Brower summarizes these contributions in a brief afterword which points to a number of further corridors to explore—and there are many.

How armed forces learn the form of modern war, and what form there is to learn, remain central questions for the training regimes of modern forces. For a number of the large armies dealt with here, the intellectual ferment concerning these questions has been most intense following national setbacks. Prussia after its humiliation at Jena, the Red Army after its defeat by the Poles in 1920, and the United States after Vietnam, all embarked on a fundamental reconsideration of how they made war.

For the United States, the harsh experience of the Vietnam War illustrated the cost of hubris, reawakening efforts to learn not only the campaign models of others but to take *mutatis mutandis* their means of perceiving the battlefield and make them their own. The revitalization of the operational art in Western military writings appears directly related to the American catharsis over the Vietnam War—a war many authorities concluded America lost. For the U.S. military, the need to generate some form of doctrine that could not so easily be subverted by the types of micromanagement and the piecemeal approach to war adopted by the Johnson administration was paramount. The American army as an institution was gravely ravaged by the travails of the Vietnam era; morale, discipline, and preparedness were all at extremely low levels. The operational doctrine first promulgated with FM 100-5 (1976) helped revitalize and refocus the army.

Finding a model of training and preparation for continuous operations, preparing to and actually keeping the enemy off balance, and striking with such rapidity that balance is never regained have been the objectives of much of the

American literature. The U.S. Army has been rebuilt since Vietnam to meet those ends—as demonstrated in the Gulf War.

Beyond revivification, the resulting doctrine packaged army methodology into indivisible form. It bears suggestion that the operational art has a political as well as a military rationale. The paradigm of the operational level clearly holds budgetary and force utilization implications. Potentially the concept serves as a vehicle for military leaders to tie the hands of those they are supposedly serving.[7] For instance, should a president desire to make war, he would be presented with army plans based on full integration of all arms. Unlike Vietnam, where the White House attempted to prevent full divisional assets from being deployed into a supposedly low-intensity theater, the new doctrine provides a ready-mixed solution that defense intellectuals, diplomats, and politicians would find far harder to disassemble. American deployments to the Gulf War serve as a clear illustration of this logic. One may or may not view this as a pernicious or perfidious formula. It is clearly a formula that solved many problems for the U.S. Army. Other armies may find the ready-mix impossible to achieve or unsuitable to their needs.

The leading powers of the Western alliance have taken an intellectual step which other members of the alliance now feel compelled to understand and follow as best they can. Most Western armed forces today recognize in training and doctrine the operational level of war. This is true of the world's largest armies: those of the United States, Russia, Britain, and Germany. Increasingly it is true of smaller nations. Perhaps incapable of waging war at the operational level themselves, these smaller powers, for instance other NATO members, have been compelled to prepare training and doctrine commensurate with their larger allies: none may ever be committed to the dance, but they must all know the steps. For these lesser powers to remain credible allies, capable of contributing to the "first team," they must come to terms with the American conception of the operational art. Many of these allies may simply adopt the American model as their own, however suitable to their own experience and needs.

The necessity of framing discussion within a lexicon that includes the operational art will continue. In a new departure, the United States has adopted the term "operational art" for its air and sea forces. The efficacy of extending a concept born on the battlefields and military academies of nineteenth-century Europe to the demands of the twenty-first century has remained largely unexamined.

Those who must ride the current trend of Western military thought outside of the United States may overlook such consideration to their detriment. Despite such concerns, the question of the operational art and its military and political influence is one of practical concern to all military policymakers, to military historians, and to the strategic studies communities. The desire evinced by several contributors to this book, among others, to develop an operational art for specialized operations such as peacekeeping tasks under United Nations auspices may prove a naive demand because it ignores the limitations and origins of the concept. However, it would not be the first time a discrete military terminology was employed well beyond its original scope and meaning. This volume provides a reminder of the origins of the concept, and may serve to focus discussion and research on its historical context. The essential historical case studies presented here will assist in making comprehensible the lively and continuing debates over the meaning of the operational art.

NOTES

1. Comments made on St. Helena, cited in Cyril Falls, *The Art of War From the Age of Napoleon to the Present Day* (Oxford, 1961), 231.

2. Jacob W. Kipp, "Lenin and Clausewitz: The Militarization of Marxism," *Military Affairs* XLIX, # 4 (December 1985), 184-91.

3. V. Ye. Savkin, *The Basic Principles of Operational Art and Tactics (A Soviet View)*, Trans. USAF. (Moscow, 1972; USGPO, 1985). Editor's preface.

4. Jacob W. Kipp, "General-Major A. A. Svechin and Modern War: Military History and Military Theory," introductory essay in Kent Lee, ed., A. A. Svechin, *Strategy* (Minneapolis, 1992).

5. Walter Goerlitz, *History of the German General Staff* (Boulder, CO, and London, 1985); Martin Van Creveld, *Command in War* (Cambridge, MA, 1985), 109-15.

6. Van Creveld, ibid., 105.

7. As suggested of the German General Staff by Williamson Murray, "What the Germans Got Right," in Michael I. Handel, ed., *Clausewitz and Modern Strategy* (London, 1989), 268.

2

The Operational Art: Developments in the Theories of War

John English

This chapter endeavours to examine the etiology and intellectual foundations of the operational art, roughly defined as pertaining to that gray area between strategy[1] and tactics. To be more definitive, if *strategy* is the art of war and *tactics* the art of battle, then *operations* is the art of campaigning. As the roots of this subject are deeply Eurocentric and military, as opposed to naval, the conduct of war at sea hardly enters the following discussion. Actually, the military concept of an operational level of war lying somewhere below strategy but above tactics is not entirely new. Indeed, the argument has been made that Napoleon Bonaparte fathered this third stratum of war through the masterful maneuver of numerous corps formations on a grand scale.[2] His prominent interpreter, General Antoine Jomini, even coined the term "grand tactics" in his 1837 *Précis de l'Art de la Guerre* to describe the art of massing troop combinations against decisive points in enemy dispositions before and during battle. One has to bear in mind, however, that Jomini emphasized making such deployments "according to the characteristics of the ground . . . in contradistinction to planning upon a map."[3] He defined strategy, on the other hand, as "the art of making war upon the map . . . [within] the whole theatre of operations."[4] Strategy, in short, determined where to act while grand tactics prescribed the manner of execution, including the employment of troops. To describe the connecting link between the two, Jomini used another term, "logistics," which had as much or more to do with the increasing need for operational staff planning and the efficient movement of large formations as with the mere provision of materiel.[5] Perhaps it should surprise no one, then, that the first proper general staff sprang from the quartermaster corps.

Although Jomini's great contemporary, Carl von Clausewitz, referred in his *On War* only to tactics and strategy[6]—defining tactics as the use of armed forces in the engagement and strategy as the use of engagements for the object of war—he patently perceived gradations of strategy. His description of "war" as "the next stage" of strategy beyond the "campaign"[7] and his later reference to the "theory of major operations (strategy as it is called)"[8] seems to indicate, moreover, that much of what he termed strategy was in fact the operational art.[9] His equation of "war, campaign and battle" to "country, theatre of operations and position"[10] and his reference to "operative elements"[11] point to a threefold division of sorts. "The strategist must," as Clausewitz elaborated, "define an aim for the entire operational side of the war that will be in accordance with its purpose. In other words, he will draft the plan of the war, and the aim will determine the series of actions intended to achieve it: he will, in fact, shape the individual campaigns and, within these, decide on the individual engagements."[12] Obviously, Clausewitz's understanding of the implications of time and space on maneuver, his grasp of maintenance and movement factors, and his insistence that the higher the level the more intellectual the challenge readily confirm that he possessed a sound comprehension of the links between tactics and strategy.[13] The moral, perhaps, is that just because a level of war lacks a label, one cannot necessarily conclude that it does not exist.

As near as can be gleaned, Helmuth von Moltke was the first to frequently employ the term *operativ* or "operational." To the pragmatic and undogmatic Moltke, strategy was little more than a system of temporary expedients to be practically applied under continually changing and challenging circumstances. Like Clausewitz, he also considered study and inquiry to be more reliable guides to success in war than any doctrine based on immutable principles or axioms.[14] According to the German historian Roland G. Foerster, Moltke adumbrated the relationship of strategy, operations, and tactics "for the first and last time" in his 1871 work, *Ueber Strategie*.[15] Noting that strategy aimed at achieving the highest politically determined objective, he described it as involving both the assembly and use of military resources in operations. The will of the enemy was, in turn, to be broken by tactics. Yet, while Moltke clearly perceived of operations as subordinate to strategy, he did not apparently envision any formal third level of war leading "an equal and individual existence next to strategy and tactics."[16] In fact, Foerster's detailed study of Moltke's scattered writings revealed that he almost exclusively used the term "operations" in the sense of the *movement* of bodies of troops for the purpose of combining forces for decisive battle. It remained for subsequent systematizers such as those in the historical branch of the German General Staff to consolidate the concept of an operational level of war within German doctrine.[17]

The times seemed ripe to do so, of course, for as the Napoleonic Wars had shown, there were definite limits to the size of an army, however well drilled or disciplined, that could be controlled by a man on a white horse on a hill. Military genius alone was no longer sufficient to shore up the generalship of large forces. The solution, first instituted by the Prussians to compensate for lack of military competence among royalist appointees, was to provide field commanders with general staff advisers capable of offering expert counsel and overseeing the detailed execution of orders. What early distinguished the Prusso-German general staff adviser from others, however, was his institutionalized right to participate in the operational decision-making process. An equally significant advantage conferred by the general staff collective was that it enabled armies to study war in peacetime and devise a "doctrine for war fighting" that would otherwise take too long to fashion during the event.[18] The core of Moltke's operational thinking, for example, included meticulous deployment planning, seeking the destruction of the enemy's army as the operational objective, granting maximum freedom of action to subordinates, and forming a center of effort to effect large envelopments and encirclements.[19] To Michael Howard, who observed that the transformation of warfare occurred before the transformation of technology,[20] the development of the German general staff system represented the greatest military innovation of the nineteenth century.[21] This was when staff colleges superseded military institutions like the Royal Military College and West Point as centers of military knowledge.

The technological advance that most affected strategy and operations during this same period was, of course, the railway. The railway made mass armies practical. Its carrying capacity not only permitted the rapid deployment of troops and horses in unprecedented numbers, but also by eliminating the need for exhausting marches ensured their arrival in good order and fresh condition. Equally important, mass conscript armies could now be sustained in the field as never before—logistically, medically, and spiritually—by rail communications.[22] In 1834 the German economist Friedrich List argued that Prussia's weak central position could be transformed into a defensive bastion through the judicious use of railways along natural interior lines, thereby raising the nation from a position of secondary military power to first rank. Thenceforth the Prussians were in the forefront of developing the military potential of trains during the railway boom that took place between 1840 and 1870. In 1846, the year List died, the successful experimental movement of a 12,000-man Prussian corps by rail convinced the Prussian General Staff to make a comprehensive survey of the military applications of railways. After the 1859 Northern Italian War, in which the French first demonstrated the military potential of large-scale rail deployments, the Prussians established a general staff railway section, the first of its kind.[23] Yet, it was not the existence of railways but, rather, the staff efficiency with which they were used that made the difference in war. The French railways were as good as the German in 1870, but the French army, concerned mainly with grand strategy, made no detailed plans for moving troops by rail. The Prussians, on the other hand, had learned from their mistakes against the Austrians in 1866 and made mobilization and rail movement timetables the centerpiece of their operational planning.[24]

The highly successful "cabinet" wars of German unification confirmed the worth of the Prussian General Staff by demonstrating that the imperatives of troop mobilization, railway movement, and army logistics could be left neither to chance nor to amateurs. Moltke's operational approach had, in fact, completely transformed the nature of war preparation and military planning. The great tragedy, however, was that while Moltke readily accepted Clausewitz's rationale that the object of war was to achieve a satisfactory political result, he categorically denied the primacy of policy by insisting that soldiers alone control the actual direction and conduct of war.[25] The crowning success of Moltke's great encirclement battles thus subverted the essence of Clausewitzian theory by lending credence to the belief that the conduct of war was best left to generals. Indeed, even though Moltke always ultimately accepted the authority of Chancellor Otto von Bismarck, the "cabinet wars" appear to have convinced most German soldiers that the relationship between statesmen and military leaders had changed since the days of Clausewitz. Count Alfred von Schlieffen, successor to Moltke and self-professed disciple of Clausewitz, especially agreed with his predecessor that military strategy served policy best by acting independently of it. Driven by fears that Germany might have to fight on two fronts, Schlieffen also enshrined the concept of the battle of annihilation in his influential work, *Cannae*, which prescribed offensive movement and concentric flank attack as the ideal method by which to produce a quick decision in a short war.[26] The development of his more famous Schlieffen Plan variant, in turn, totally ignored Clausewitz's counsel that no major plan of war should be based solely on purely military considerations. By the summer of 1914 the policy of the Second Reich largely reflected the dictates of an offensive war plan. German politicians, moreover, considered it their duty to shape political measures to accommodate this plan.[27]

The suggestion has been made, of course, that the German General Staff may have introduced the operational sphere less to ride the rails than to proscribe political interference in military operations.[28] Whatever the case, it seems clear that German strategic thought devolved downward, toward the tactical and operational levels, rather than upward. German grand strategy was ultimately a military strategy[29] that, according to Colmar von der Goltz, "concerned itself with those large-scale measures which serve to bring the forces into play at the decisive point under the most favourable conditions possible."[30] Those who succeeded Moltke increasingly identified strategy with operations to the point of denying political and diplomatic factors any influence over military matters even in peacetime. Although the Germans produced six successive editions of *On War* in the course of the First World War, most of their officers continued to reject Clausewitz's postulates on the primacy of politics and the superiority of the defensive. They also preferred to read the more practical and prescriptive doctrine of Schlieffen presented in *Cannae* rather than the more difficult and contemplative discourses offered in *On War*.[31] Significantly, the First World War also witnessed the development of the German army "chief system" in which highly competent first general staff officers actually commanded armies led by, presumably, less capable nobles and princes.[32] Within Germany itself, in the meantime, the military simply subsumed German civil government, but with no accompanying gain in military power.[33]

As much as one might criticize the Germans for their abiding neglect of the grand strategic level, however, they proved to be pacesetters in the operational and tactical realms. The introduction of the "elastic defense in depth" and the development of "storm troop tactics," in particular, changed the face of battle for all time. The difficulty was that the basic problem confronting most First World War combatants was tactical in nature: how to weather the storm of steel across "the last 300 yards."[34] More often than not, large-scale operations foundered on the inadequacy of tactics to negotiate this distance. Even the highly innovative offensives launched by the Germans in March 1918 ultimately failed, in large part, because of their inability to sustain the necessary tactical standards.[35] In the end, they were defeated operationally by the railway, which by enabling the defending Allies to deploy reserves faster than the attacking Germans, proved to be a more decisive factor than any weapons system in winning the First World War. Ironically, earlier German fears about rising Tsarist railway construction enhancing Russian mobilization capacity may also have prompted the fatal German decision to risk war in the first place.[36]

By almost any military measure the First World War can only be described as a historical watershed that marked the beginning of modern warfare. The extensive use made of railways, without which the First World War could not possibly have been fought in the manner it was, plus the unparalleled level of staff coordination required to handle unprecedented numbers of troops and materiel marked it as a war unlike others. Of all modern military advances wrought by the First World War, however, none was more significant than the development of the indirect fire capabilities of artillery. Indeed, the widespread application of indirect fire was alone enough to set the Great War apart from all previous wars that, henceforth, could only be regarded as old-fashioned affairs in which the field gun performed a role more akin to that of the tank than modern artillery. Certainly, the mostly smoothbore cannon trotted out in the U.S. Civil War and man-handled through such places as the thickets of Chickamauga were little more than direct fire platforms. In the Great War, by way of contrast, neither the machine gun nor the "machine gun destroyer" that came to be called the tank inflicted as many casualties as the "steel rain" of artillery. Whereas before the Great War small arms fire inflicted between 80 and 90 per cent of battle casualties, shell fire thereafter accounted for more than 60 per cent.[37] When coupled with improvements in military medical services, this distinctly modern phenomenon also ensured that for the first time in warfare more deaths resulted from combat than from disease.[38]

Obviously, too, the sheer magnitude of the logistical dimension associated with the provision of supplies and war *materiel* for eventually more than 400 divisions on both sides[39] contributed to this sea change in warfare. Whereas ammunition constituted but 1 percent of total field supply as late as 1870, it increased roughly tenfold after 1914. A British division required about 27 wagon loads of supplies per day in 1914; two years later, daily consumption of such items as food and fodder still stood at 20 wagons, but the number required to carry combat materiel, including ammunition, had risen to around 30.[40] Decisions concerning the allocation of rounds per gun, approved rates of daily expenditure, and the selection of dumping locations thus became increasingly important operationally. Determining the optimum allotment of guns and ammunition for tasks related to covering troop movement and counterbattery fire, the latter almost exclusively an artillery/air[41] matter, similarly called for gunner expertise and detailed staff coordination. The consequent requirement for the centralization of all artillery resources in order to use them more efficiently, however, appeared to run counter to the growing need for greater tactical flexibility at lower levels. The obvious solution, of course, was to include artillery commanders in the planning process at all levels. More than ever before, success in battle depended upon the *orchestration*[42] of arms, not simply in the spatial maneuver of forces, but in the timely coordination of the fall of shot with movement.[43] The shell, not the gun, was the weapon of the artillery, and fortified defenses protected by indirect fire could only be taken by troops supported by artillery. The advent of the tank, which maneuvered spatially like the infantry, was of far less import from the standpoint of synchronization. At the same time, the introduction of the wireless-equipped aircraft, initially to spot and photograph enemy dispositions for artillery and, later, to carry out tactical attack missions, expanded the range and means of projecting indirect fire as well as the size of the all arms team.[44]

Nowhere was the modernity of the First World War better illustrated than during the battles of the "Hundred Days," which started with the attack at Amiens on 8 August 1918 and finished with the capture of Mons 100 miles away on Armistice Day. At Amiens, interarm cooperation among the forces of the British Empire reached new heights as aircraft, guns, tanks, and infantry all acted in concert. Tanks and infantry assaulted under the cover of a fast-moving predicted barrage and a hurricane counterbattery bombardment. While loitering spotter planes protected by fighters directed friendly artillery fire, other aircraft engaged enemy troops more directly in a ground attack role,[45] proving particularly effective against enemy antitank guns. Wireless central information cells established at corps level coordinated the bulk of these air and artillery activities and acted as the hubs of the mobile battle. The principle that the air battle had first to be won before the artillery and then the maneuver arms could succeed also came to be accepted. Although wireless communications were not fitted into tanks and only extended as far forward as brigade level, over 160 attacks involving tanks were launched during the Hundred Days. Indeed, the system of all arms integration used by British imperial forces during this period clearly presaged the shape of things to come in the Second World War.[46]

Unfortunately, the lessons of the Hundred Days series of battles escaped official notice in the rush to demobilize, though the Germans who continued to focus on the operational level of war apparently studied the Battle of Amiens.[47] The legacy of the First World War left enough of an imprint nonetheless to encourage some progressive British military thought on the subject of large-scale operations. In his *Foundations of the Science of War*, Colonel J. F. C. Fuller expounded upon "grand tactics" to a considerable extent.[48] If the "correlation of the forces of war [to the political object was] the main duty of the grand strategist," he wrote, "the duty of the grand tactician" was to endow those forces with a structure and employ them in a campaign. To Fuller, grand strategy secured the political object by directing all moral, physical, and material resources toward the winning of a war, while grand tactics involved taking "military action by converging all means of waging war towards gaining a decision." Rather interestingly, Fuller saw the grand tactical object as the "destruction of the enemy's plan" so as to reduce his will to win and thereby force his surrender or make him sue for peace. In postulating that it would be "an error of the first magnitude" for the grand tactician to think solely in terms of physical destruction, usually the aim of minor tactics,[49] he evinced a degree of sophistication not always apparent at this level of war.

It would be foolish to suggest, of course, that Fuller had any corner on the subject of "grand tactics" during the interwar years. One suspects, for example, that the U.S. Army's Lieutenant Colonel Charles Willoughby, who studied at the University of Heidelberg and instructed at the Command and General Staff School, merely reflected a common view when he noted in his *Maneuver in War* that "military thought is continually troubled by an apparent conflict in the tactical concepts of 'large' and 'small' units or a twilight zone between them."[50] For the most part, the Command and General Staff School taught theater operations as "military strategy."[51] In his 1920 work, *Generalship in the World War*, Baron Hugo von Freitag-Loringhoven also observed that in the German army the term "strategical" had generally fallen into disuse and been replaced by "operations" to "define more simply and clearly the difference from everything that is referred to as 'tactical.'" He went on to explain that "operations" as a rule embraced what took place "independently of actual combat," while "strategy" pertained "to the most important measures of high command."[52] The aberrant breakthrough of Panzergruppe Kleist on the French front during four days in May 1940 could therefore aptly be hailed as "the first time armour was used in the operational role."[53] But here again, the key was *movement*, which the Germans facilitated by establishing major fuel depots close to the border and by having the Panzergruppe carry the bulk of its supplies. Trouble only arose because that formation, with over 41,000 vehicles at its disposal, was allocated but four routes.[54] Such faulty operational planning was nonetheless "put right by the tactical mobility of the middle and lower command."[55]

Notwithstanding the operational flair exhibited by the Germans to this point, however, the intellectual underpinnings of their military thought at this level may not have been as advanced as those developed earlier by Soviet theorists. In fact, A. A. Svechin of the Frunze and General Staff Academies was probably the first to propose, in his 1927 work *Strategy*, the concept of operational art as a new and distinct category of military theory. According to Svechin, decisions could no longer be attained through single engagements. Tactics thus provided the "steps" for operational "leaps" along the path determined by strategy. By 1933 the Red Army had also officially sanctioned as tenets of Soviet military art the concept of deep battle and, after the publication of the *Field Regulations* of

1936, prepared under Marshal M. N. Tukhachevsky, the concept of deep operations. Unfortunately, the innovative military theorists who first formulated the ideas of operational art and deep battle died in the Stalinist purges of 1937-1938. The Soviets in the initial period of the Second World War thus found themselves having to relearn the hard way the advantages of such military thought.[56] Until they stopped German offensive action at the tactical level, they had no hope of succeeding strategically and operationally. By the time of the Battle of Kursk, where the Germans for the first time failed tactically to break into Soviet operational depth, the Soviets had, however, fashioned a truly formidable operational offensive capability that forced the Germans to relearn how to conduct First World War "elastic" defense.[57] But it was too late, as the Russians swept all before them in a series of major thrusts. In Operation Bagration, launched on 22 June 1944, the anniversary of the Hitlerian and Napoleonic invasions of Russia, four Soviet fronts eliminated about 28 German divisions in three intentionally limited encirclements over 600 kilometers.[58] In the words of one panzer staff officer on the receiving end, the Russians "simply crushed Army Group Centre and battered it to death."[59] Similarly, in the Vistula-Oder operation of January-February 1945, the Red Army advanced an additional 600 kilometers on a front of 500 kilometers in three weeks, annihilating 35 German divisions and inflicting 50-75 percent casualties on a further 25.[60]

According to Sandhurst Sovietologist Charles Dick, the spectacular Russian victories of 1943-1945 were decidedly not the product of brute force and ignorance as long thought in the West. They instead reflected the application of a highly refined operational art that aimed at the disruption of an enemy's cohesion on a large scale, thus depriving him of the ability to react to changes in the situation, breaking up his organization and control of higher formations, and, ultimately, preventing him from accomplishing his aims. While this necessarily restricted the latitude of tactical commanders who received detailed orders, the Russians at the operational level displayed considerable skill in being able to deceive the enemy by covertly maneuvering and massing combined arms armies on breakthrough axes, and, subsequently, launching mobile groups of armor,[61] supported by fleets of air armies, to cut through to unprecedented operational depth. Although the Germans continued to score tactical victories, they were utterly outmaneuvered on the operational level, where the rate of advance for Soviet tank armies reached an average of 20-50 kilometers a day in 1944-1945. If there was any lesson from the mighty struggle on the Eastern Front, it was that winning a battle was of little consolation if one lost the campaign.[62] In the postwar period the Soviets never forgot this lesson, even though during the period of "nuclear nervousness" up to the mid-1960s the operational art received comparatively less attention than strategy.[63] Significantly, however, the attention paid by the Soviets thereafter to the operational level of war did not attract the interest of NATO authorities until the 1980s.[64]

No doubt the essentially nuclear stance of NATO had something to do with this, as well as the war-fighting experience of the Western Allies during the Second World War. Neither the NATO nor the Western Allied armies approached in size that of the Soviet Union, which almost singlehandedly brought about the defeat of the Third Reich. The Second World War was won on the Eastern Front, just as the First World War was won on the Western Front, and both with a lot of hard pounding. A major difference was that the technological brake upon field mobility which characterized First World War operations no longer existed in the Second World War. At the same time, troop densities in the comparatively more restricted Mediterranean and northwest European theaters did not always allow the same scope for ground force maneuver as did the Eastern Front.[65] Be that as it may, the Western Allies still managed to bottle up roughly 250,000 Axis soldiers in Tunisia and, by John Keegan's[66] calculations, grind to bits 27 German infantry and 11 panzer divisions in the Normandy campaign. The only encirclement carried out was that at Falaise, which bagged 50,000 German prisoners and begged the question of whether Western Allied generals knew anything about the operational art. In the controversy that ensued, however, it may well have been historians who did not understand, for with their eyes firmly fixed on the shorter envelopment at Falaise they failed to see that the advance to the Seine ten days ahead of schedule was of far greater import from an operational perspective.[67] The subsequent lightning drives by the armies of Generals George Patton and Miles Dempsey toward Lorraine and Brussels, the last covering 200 miles in six days, additionally tend to belie any endemic ignorance among the Western Allies of operations above corps, which according to the doctrine of the time was at least recognized as the highest tactical formation.

This is not to suggest that the Western Allied commanders who led their forces ashore in Europe from 1943 onward possessed as profound an understanding of the operational level of war as did their more experienced German and Russian counterparts. We know, however, that despite the lack of such precise terminology, many of them had a better grasp of how to maneuver armies and army groups than armchair generals give them credit for. For example, when Patton learned on 23 August 1944 that General B. L. Montgomery was most likely to be authorized to mount the major Allied thrust on Germany along the axis north of Paris, he "thought up the best strategical idea he ever had," which was, briefly, to strike north to Beauvais and, by paralleling the Seine, move west to open it for the British and Canadians.[68] It had all the markings of a brilliant stroke. Similarly, Operation Grenade, which threw General William Simpson's Ninth U.S. Army against the exposed flank and rear of German positions forward of the Rhine, represented a "classic example of skill in manoeuvre."[69] There can be no question, as well, but that the Western Allies excelled to a greater degree in the area of higher strategy, which witnessed the establishment of a matchless Combined Chiefs of Staff system that worked in concert with far-flung geographical theaters of war under supreme commanders. In an associated vein, Western Allied naval and ground force coordination reached new heights of efficiency that would have made Julian Corbett proud, particularly in Europe, where the greatest amphibious operations were conducted, mainly by the venerable Royal Navy.[70] The success with which the war machinery of the Grand Alliance functioned, in turn, set the tone for the military infrastructure of NATO.

Contemporary American and British interest in the operational level of war and the activity known as operational art, of course, only dates back to the 1970s when the U.S. Army sparked a renaissance in military thinking in the aftermath of the Vietnam War. A troubling concern at the time was that the tactical performance of troops might not even matter if strategy and grand tactics proved faulty. Another challenge was to be able to fight outnumbered and still win a major land campaign in Europe. The publication in 1976 of Field Manual (FM) 100-5, *Operations*, which concentrated solely upon war fighting, represented a first step toward addressing this situation. A new manual, FM 100-1, *The Army,* later incorporated Clausewitzian theory into strategic doctrine and lent renewed emphasis to Fuller's principles of war first adopted by the U.S. Army in 1921. Although FM 100-5, *Operations*, reflected the lessons of 1,000 battles, including those of the 1973 Yom Kippur War,[71] it was severely criticized for emphasizing an "active" defense based largely on the employment of weapons systems in a series of delaying actions. The subsequent publication of FM 100-5, *Operations* (1982), however, proclaimed the doctrine of AirLand Battle and the associated concept of the operational art,[72] while stressing the principles of war[73] and maneuver as the dynamic element in combat. FM 100-5, *Operations* (1986), reinforced the idea of integrating all war-fighting resources within an operational theater and waging simultaneous battles in the forward, deep, and rear areas of the enemy. The 1986 version of FM 100-5 also repeated Clausewitz's dictum that the whole of military activity should relate to the engagement.[74]

The perception that maneuver at the operational level could make up for numbers no doubt propelled the U.S. Army in the direction that it took. This fanned, coincidentally, a maneuver warfare school of thought that may have first originated at the tactical level in the U.S. Marine Corps and spread rapidly out from there even into the hallowed halls of the British Army Staff College Camberley. As explained by one proponent, the idea of maneuver in the sense of maneuver warfare was best defined by the theory developed by Lieutenant Colonel John Boyd of the U.S. Air Force From an examination of mock air-to-air combat exercises at Nellis Air Force base in 1974, and subsequent historical studies of Korean "dogfights" and various land battles, Boyd deduced that the advantage invariably went to the side that most rapidly completed what he termed the "observation-orientation-decision-action (OODA) cycle." The slower the reaction to the "OODA Loop," the greater the risk of disaster, as befell the Romans at Cannae and the French at Sedan. Besides postulating that only a decentralized military could have fast OODA Loops, the maneuverist school advocated the acceptance of confusion and disorder on the battlefield, so as to be able to operate better within it, and the avoidance of all predictable patterns of action. The aim was to fight smart, employing as required mission-type orders, the *Schwerpunkt* principle (supporting the successful wing in an attack), and the concept of *Flaechen und Luekentaktik* (tactics of positions and gaps) to effect a sort of "military judo" on an enemy. Although the maneuverist school focused heavily on tactics, it also offered that "excellence in the operational art more than . . . manoeuvre in tactical battle" enabled "a smaller force . . . [to] defeat a large one."[75]

In Britain, the military intellectual ferment occasioned by the publication of FM 100-5 (1982) ultimately led the British Army to incorporate the operational level of war into its doctrine in 1989.[76] Perhaps the most prolific, if not the most influential, British military thinker to address this issue was the late Brigadier Richard Simpkin, a latter-day Fuller, who liked the idea of smaller forces using an opponent's own mass and momentum to overthrow him. To Simpkin, "operational" meant "having the dynamic characteristics associated with manoeuvre theory."[77] He also argued that technological advances coupled with maneuver theory had lowered the threshold of what previously constituted operational level. Any Soviet airborne assault brigade that seized a Rhine bridge intact, for example, would have been an operational formation by virtue of the fact that its act affected the whole theater.[78] Presumably Lawrence of Arabia's inspired refusal to take Medina during the Arab Revolt, in order to attack Turkish troop movements along its one and only rail link, fitted this same category. "So far from being concerned only with large masses," wrote Simpkin, "manoeuvre theory is in fact a formal statement of the kind of warfare historically conducted by small organized forces of high quality."[79] In stating the case for maneuver warfare,[80] Simpkin, like most enthusiasts, also employed the so-called attrition model of warfare which has not always reflected historical reality.[81] To his great credit, however, he went on to admit that once "inelegant fighting" occurred, "the two theories become complementary."[82] He nonetheless held, as did German General Franz Uhle-Wettler, that it was possible to throw tactical monkey wrenches into the component parts of Soviet operational machinery by using a combination of conventional and quasi-guerrilla fighting techniques.[83]

Within the Soviet camp, meanwhile, the revision of war-fighting concepts first undertaken in the mid-1970s picked up momentum after Marshal N. V. Ogarkov became chief of the General Staff in 1977. In reviewing the growing size of armies since the 1500s, Ogarkov came to the conclusion that increasing numerical strength had progressively increased the spatial scope of military actions. The formidable accuracy and extended ranges of emerging conventional weaponry also convinced him that older forms of front or army group operations had to be replaced by theater-level "strategic operations" if desired political-military objectives were to be achieved. By integrating long-range, terminally guided missile systems into automated reconnaissance strike complexes, the whole territory of a country could be attacked. Given the enhanced mobility of armored forces, attack helicopters, and airmobile formations, Ogarkov envisioned operations being conducted over long distances at high speeds by combined arms formations. In this context, advance guards, forward detachments, and higher-level Operational Maneuver Groups (OMGs) resembling the mechanized and tank corps of the 1930s were expected to play a considerable role. The enlarged depth envisaged for future conflict, in fact, reflected the earlier theoretical concepts of deep battle espoused by Tukhachevsky and other Tsarist-trained thinkers, such as V. K. Triandafillov.[84]

When Ogarkov first officially announced the expansion of the scope of operations in 1982, he also upgraded the importance of the initial nonnuclear phase in future war. His view, supported by Secretary Leonid Brezhnev's public pledge of 15 June 1982 that the Soviet Union would not be the first to use nuclear weapons, was that it was criminal to look upon nuclear war as a rational act. Convinced that rough nuclear parity meant that nuclear weapons had negated themselves in terms of military utility, Ogarkov had even suggested, in 1979, that a world war might begin and end conventionally. Theater operations could be conducted without reliance on nuclear systems, moreover, since enemy field forces could be destroyed more efficiently through the employment of technologically advanced weaponry firing precision-guided and improved conventional munitions. The ability of higher commanders to deliver air and missile strikes deep in enemy depth seemed also to confirm the importance of operational art over tactics. A defender armed with such military technology could also conceivably wrest the initiative from an attacker by striking back quickly throughout the latter's depth. This development, in the estimation of Ogarkov, amounted to nothing short of a military revolution and, henceforth, was likely to be the single most decisive factor in combat.[85] As things turned out, of course, such thinking was eventually overtaken by events.

In the interim, NATO planners had continued to examine how to deal with a possible offensive by several Warsaw Pact *echelon fronts*, each comprising first and second *operational echelons*, which in turn would comprise first and second *tactical echelons*.[86] One result was the formal sanctioning, in 1984, of a theater level operational subconcept termed "follow-on-forces attack" (FOFA). Reputedly the brainchild of General Bernard Rogers, who recognized that operational depth was politically difficult to attain within Germany, FOFA cleverly looked to gain it on the enemy side. While Rogers was reasonably confident that NATO corps could handle the first operational echelon, he feared that subsequent second echelon forces and OMGs would prove too much for them. In such circumstances, which translated into but a few days of war, he felt he would have little choice but to request nuclear release sooner rather than later. He reasoned that NATO forward defenses could be made to work without resort to nuclear fire, however, if rearward Warsaw Pact forces could be simultaneously disrupted and destroyed by conventional means up to 300 kilometers in their depth. This was to be accomplished by FOFA through the application of proven technologies such as manned aircraft, missiles with conventional warheads, remotely detonated mines, cratering munitions, and electronic warfare. Although FOFA included a naval component designed to interdict Warsaw Pact amphibious vessels, it did not involve airmobile or other ground forces in cross-border operations.[87]

What would actually have happened had war broken out between NATO and the Warsaw Pact remains a matter for speculation. Perhaps the application of the operational art by both sides saved Europe and more from a fate much worse than Chernobyl. On the other hand, FOFA may not have worked and NATO corps would have been left to bear the brunt of battle themselves, hoping, as Simpkin suspected, that tactical successes here and there would unhinge the Soviet operational machine as happened to the 1918 German offensive in the First World War. The NATO corps themselves, at least those in Central Army Group protected by huge barrier minefields, would not have maneuvered as corps, since ground and size would have precluded them from doing so efficiently. Only divisions would have been able to maneuver, with the possible exception of the First French Army, whose 150,000 men and 1,000 tanks could have counterattacked as far as the River Main, where logistical requirements would have forced them to stop.[88] To those who would suggest that NATO proved its war-fighting superiority in the Gulf, it warrants counter that the grand operational victory scored there was accomplished against a second- and perhaps even third-rate enemy who, lacking sufficient powers of observation, could fight only like a blind man. This is not to suggest that the Gulf victory was anything less than a remarkable feat, however. Quite clearly, neither France nor Britain, still powerful nations in their own right, could have taken on Iraq without resorting to major and highly disruptive mobilization. So here again is part of the modern answer to the age-old question of critical mass: what happens when a small, highly mobile, well-led army meets a large, highly mobile one, perhaps not so well led?

If anything, the staggering *logistical* and *staff planning* requirements of the Gulf War should serve as reminder that it is indeed these dimensions as much as sweeping battlefield maneuvers that characterize the operational art, just as Jomini intimated so many years ago. Yet, with due respect to Jomini, one could argue that the dawn of the operational level of war really only broke when general staff officers started to load masses of troops on trains. The mundane business of *movement* calculation, moreover, seems still to lie at the heart of this level of war, much as it did when operations first began to ride the rails. It remained for the Soviets, however, to truly formalize in theory and practice the concept of the operational art as a distinct level of warfare in terms of mission, scale, scope, and duration.[89] Perhaps we should not be too surprised at this, for, in fielding the largest armies of the century, the Soviet Union and Germany were also the first to encounter challenges associated with the effective employment of those armies. The Western Allies and NATO powers, in contrast, remained more oceanic in orientation and hence more strategic in outlook, which perspective in the form of "military strategy" sufficed for armies landed by sea and operating on littorals as opposed to steppes. After confronting the most powerful land force in the world for half a century on the Central Front, however, there were many in the NATO camp who grew to admire the sophistication of the Soviet war-fighting approach. Therefore, the adoption of the operational level of war by the Anglo-American powers was probably an imitative as well as progressive step, even if, in the British case, it occurred in the very same year that the satellite states imploded.

One could almost say that NATO learned more from the Soviets than they learned from the West. Obviously, there are many aspects of the operational level of war that deserve serious study, among which are, to name but a few, the timely commitment and creation of reserves, the use of river lines, the correct handling of salients, the nuances of parallel pursuit, and the advantages of encircling versus leaving escape gaps. Of course, there are always dangers associated with the inflexible compartmentalization of war, not the least being the proprietorial tendency of those working at one level to decry the importance of other levels. For soldiers to focus exclusively on the operational art to the detriment of tactics, while leaving the field of strategy to civilian academics, would similarly be unwise. As for the expectation that supporting objectives will cascade with logical precision from war aims, it might be well to recall that British politicians refused even to permit the rigorous debate of war aims during the First World War.[90] Neither have strategic goals always been easy to select; Passchendaele had a strategic goal, but it seemed to get lost in the struggle for a ridge. The Scheldt in the Second World War was also arguably a strategic objective, but taking it ultimately depended more on minor tactics than any operational-level action. Clearly, there will be times in war when armies will not be able to maneuver (just as there will be times when it will be prudent to wait),[91] but the essential unity of strategy, operations, and tactics will remain. The larger question is whether their differences will blur entirely as highly automated air and space systems come more and more into play.[92] Perhaps, because war is more of a social than technological phenomenon, the answer will lie somewhere between the Gulf and Grozny. Given that the operational art originally sprang from the maneuver of large formations, it also remains to be seen whether it can be profitably applied by small armies in pursuit of strategic objectives. To attempt to relate the concept to everything from internal security to peacekeeping, drug wars, and more may only invite muddle.

To keep the operational art institutionally alive as a war-fighting concept, on the other hand, would show good long-term judgment, as the days of large armies and great wars just might not be over. As one British officer recently pleaded: "[W]e must not settle back to thinking small, no matter how reduced our army is to become," but "this will not be easy to avoid, for we positively enjoy small wars."[93] Yet the challenge will be to avoid turning something that could truly enhance military intellectual depth into a mindless cult of the arcane. All too often, E. M. Forster wrote in *A Passage to India*, "soldiers put one thing straight, but leave a dozen crooked."[94]

NOTES

1. Which, although translated from the Greek *strategos* as "the art of the general," was actually a latter-day neologism first coined by the French writer Jolly de Maizeroy, who was active in the military field during the years immediately preceding the French Revolution. Martin van Creveld, *The Transformation of War* (New York, 1991), 96.

2. Colonel Wallace P. Franz, "Grand Tactics," *Military Review* 12 (December 1981), 32, 34.

3. Antoine Henri Jomini, *Summary of the Art of War,* 115, 149.

4. Ibid., 115. Jomini considered strategy the same under Caesar as under Napoleon, but thought tactics "the only part of war, perhaps, which it is possible to subject to fixed rules." See ibid., 88.

5. Ibid., 115, 181-88; and Michael Howard, "Jomini and the Classical Tradition in Military Thought," in Michael Howard, ed., *The Theory and Practice of War* (Bloomington, IN, 1965), 14-15.

6. "The distinction between tactics and strategy is now almost universal," Clausewitz wrote, "and everyone knows fairly well where each particular factor belongs without clearly understanding why." While arguing that "whenever such categories are blindly used, there must be a deep-seated reason for it," he at the same time rejected "the artificial definitions of certain writers." Carl von Clausewitz, *On War*, edited and translated by Michael Howard and Peter Paret (Princeton, NJ, 1976), 128.

7. Ibid., 358.

8. Clausewitz, "Unfinished Note, presumably written in 1830," cited in ibid., 70.

9. Wallace P. Franz, "Two Letters on Strategy: Clausewitz's Contribution to the Operational Level of War," in Michael I. Handel, ed., *Clausewitz and Modern Strategy* (London, 1989), 171-93; and his earlier seminal works "Grand Tactics," *Military Review* 12 (December 1981), 32-39, and "Maneuver: The Dynamic Element of Combat," *Military Review* 5 (May 1983), 2-12.

10. Clausewitz, *On War*, 379.

11. These included such aspects as superiority of numbers, surprise, cunning, concentration of forces in time, the strategic reserve, economy of force, and the geometrical factor. See ibid., 225.

12. Ibid., 225.

13. Ibid., 140, 280-81, 292-301, 314-47.

14. Ibid., 141.

15. Roland G. Foerster, "The Operational Thought of Moltke the Older and Its Consequence," in *The Operational Thought of Clausewitz, Moltke, Schlieffen, and Manstein* translated by Multilingual Translation Directorate, Secretary of State Canada (Freiburg, 1989), 28-29.

16. Ibid., 29-30.

17. Ibid., 30.

18. Oberst i. G. Christian O.E. Millotat, *Understanding the Prussian-German General Staff System* (Carlisle, PA, 1992), 19-21, 23, 26-31, 37, 41-43, 59-60; Michael Howard, *The Franco-Prussian War: The German Invasion of France, 1870-1871* (London, 1981), 23-29; Peter Paret, *Clausewitz and the State: The Man, His Theories, and His Times* (Princeton, NJ, 1985), 68-71, 125, 137-46, 287-88; Gordon A. Craig, *The Politics of the Prussian Army 1640-1945* (London, 1955), 45, 63, 78-79, 193-95; and Hajo Holborn, "The Prusso-German School: Moltke and the Rise of the General Staff," and Gunther E.

Rothenberg, "Moltke, Schlieffen, and the Doctrine of Strategic Envelopment," in Edward Meade Earle, ed., *Makers of Modern Strategy* (Princeton, NJ, 1960), 281-325. See also "The Staff Caste" in John A. English, ed., *The Canadian Army and the Normandy Campaign: A Study of Failure in High Command* (New York, 1991), 89-106; and Colonel T. N. Dupuy, *A Genius for War: The German Army and the General Staff, 1807-1945* (London, 1977).

19. Foerster, "The Operational Thought of Moltke," 30-45.

20. Sir Michael Howard, "Afterword — Tools of War: Concepts and Technology," in John A. Lynn, ed., *Tools of War: Instruments, Ideas, and Institutions of Warfare, 1445-1871* (Urbana, IL, 1990), 242; and Michael Howard, *War in European History* (Oxford, 1976), 102.

21. Ibid., 100.

22. Ibid., 98. The simultaneous development of the electromagnetic telegraph and the railway further strengthened the strategic value of rail lines of communication.

23. Colonel J. F. C. Fuller, *The Foundations of the Science of War* (London, 1925), 2-3, 95, 100-101, 106; Edward Meade Earle, "Adam Smith, Alexander Hamilton, Friedrich List: The Economic Foundations of Military Power," in Earle, ed., *Makers of Modern Strategy*, 148-51; and Theodore Ropp, *War in the Modern World* (New York, 1962), 161-62, 171, 184-94. During the U.S. Civil War, the first great conflict of the steam age, the railway also proved to be an indispensable adjunct to water transportation, which, it needs to be stressed, remained the principal means of supply and movement.

24. Ropp, *War*, 156-57, 161, 171; and Martin van Creveld, *Technology and War: From 2000 B.C. to the Present* (New York, 1991), 158-61.

25. Dennis E. Showalter, "Total War for Limited Objectives: An Interpretation of German Grand Strategy," in Paul Kennedy, ed., *Grand Strategies in War and Peace* (New Haven, CT, 1991), 107.

26. As Moltke accomplished at Sedan through a "second Cannae."

27. Jehuda L. Wallach, "Misperceptions of Clausewitz's *On War* by the German Military," in Handel, ed., *Clausewitz and Modern Strategy*, 220-22, 227-29; Foerster, "The Operational Thought of Moltke," 46-53; and Clausewitz, *On War*, 365, 367, 607.

28. Williamson Murray, "What the Germans Got Right," in Handel, ed., *Clausewitz and Modern Strategy*, 268.

29. Showalter, "Total War for Limited Objectives," 105-6.

30. Field Marshal Colmar von der Goltz referred to strategy as "the science of generalship" and tactics as "handling troops" in the engagement itself.

31. Wallach, "Misperceptions of Clausewitz's *On War* by the German Military," 216, 220-24, 226-32.

32. Millotat, *Understanding the Prussian-German General Staff System*, 41-43.

33. Showalter, "Total War for Limited Objectives," 114-15.

34. Ibid., 114.

35. Tim Travers, *How The War Was Won: Command and Technology in the British Army on the Western Front 1917-1918* (London, 1992), 50, 86-88, 90, 96, 99, 108, 175; and Paddy Griffith, *Battle Tactics of the Western Front: The British Army's Art of Attack 1916-18* (New Haven, CT, 1994), 8, 60, 195.

36. Holger Herwig, "Imperial Germany," in Ernest R. May, ed., *Knowing One's Enemies* (Princeton, NJ, 1986), 87. See also John Gooch, *The Prospect of War: Studies in British Defence Policy 1847-1942* (London, 1981).

37. The First World War produced more than nine million battle deaths, including roughly five million slain by "big guns" and three million killed by "small arms." In comparison, small arms accounted for about 90 percent of battle casualties in the U.S. Civil War, 84 per cent in the Austro-Prussian War, and 80 per cent in the Franco-Prussian War. According to the U.S. Army Medical Department, American forces in the First World War suffered 26 percent of all hits from small arms fire and 65 percent from shells; in the Second World War the figures were 20 percent and 60 percent respectively. George Raudzens, "Firepower Limitations in Modern Military History," *Journal of the Society for Army Historical Research* 271 (Autumn 1989), 134-35, 148, 151. Raudzens also claims that 60 percent of injuries in the Crimean War were attributable to small arms and 40 percent to artillery. See ibid., 147-48. Another estimate, which stressed the siege nature of the Crimean War, indicated that artillery caused 50 percent and infantry small arms 40 percent of casualties. Colonel T. N. Dupuy, *The Evolution of Weapons and Warfare* (London, 1980), 199. Although artillery in the Russo-Japanese War inflicted but 10 per cent of total casualties, it nonetheless demonstrated the worth of indirect fire.

38. Of the one and four million terminal casualties respectively caused by the French Revolutionary and Napoleonic wars, between 80 and 90 percent succumbed to disease. See Quincy Wright, *A Study of War* (Chicago, 1970), 59; and Raudzens, "Firepower Limitations in Modern Military History," 135. Ten times as many British soldiers died of dysentery as from Russian weapons in the Crimea and five times as many from disease as from battle in the Boer War. Owing to improved medical services, Japanese losses from disease in the Russo-Japanese War were less than a quarter of deaths from enemy action. In the First World War, losses from disease remained well below losses from enemy action, so long as organization and discipline remained intact. This held true even on the Eastern Front during an outbreak of typhus in 1915. In the Austro-Hungarian army, despite prolonged exposure to a typhus epidemic raging in Serbia, disease losses never exceeded 50 percent of losses from enemy action. William H. McNeill, *Plagues and Peoples* (New York, 1989), 251-52, 328.

39. James L. Stokesbury, *A Short History of World War I* (New York, 1981), 289; and John A. English, *On Infantry* (New York, 1984), 12. Roughly speaking, there were 90 German divisions opposed to 140 Allied in 1915, 120 to 160 in 1916, 140 to 180 in 1917, and 197 to 220 in 1918.

40. Martin van Creveld, *Supplying War: Logistics from Wallenstein to Patton* (Cambridge, England, 1977), 110, 141. Van Creveld also notes that in the Franco-Prussian War the Germans fired an average of 199 rounds per artillery piece, but in 1914 the roughly 1,000 rounds per gun held in German stocks were depleted within a month and a half. According to John A. Lynn, ammunition accounted for 5-8 percent of U.S. ground force supply during the Second World War, food about 10 percent, fuel 16-17 percent, and construction material 11-18 percent. John A. Lynn, "The History of Logistics and Supplying War," in John A. Lynn, ed., *Feeding Mars: Logistics in Western Warfare from the Middle Ages to the Present* (Boulder, CO, 1993), 21-23.

41. By 1917 90 percent of British counterbattery observation was conducted by airmen in wireless-equipped aircraft. See Shelford Bidwell and Dominick Graham, *Fire-Power: British Army Weapons and Theories of War 1904-1945* (London, 1982), 143.

42. On the use of this word in the context of operational art, see Colonel Richard M. Swain, "The Written History of Operational Art," *Military Review* 9 (September 1990), 103.

43. To do this, it was necessary to survey the Western Front, which produced the first gridded and contour-lined maps as principal combat tools.

44. Bidwell and Graham, *Fire-Power*, 99-115, 137-45, 163-64.

45. In the Passchendaele campaign both the Germans and the British used aircraft in the close support role; the former to support their counterattack divisions and the latter to counter counterattacks. See ibid., 143.

46. Ibid., 140-45; and Griffith, *Battle Tactics of the Western Front*, 14.

47. Captain Shane Schreiber, "The Orchestra of Victory: Canadian Corps Operations in the Battles of the Hundred Days 8 August - 11 November 1918," M.A. Thesis, Royal Military College, 1995.

48. The highly influential Colonel G. F. R. Henderson, who taught at Camberley, the British Army Staff College, before Fuller, described "grand tactics" as "methods employed for . . . [enemy] destruction by a force composed of all arms." See G. F. R. Henderson, *The Science of War* (London, 1919), 71.

49. Fuller, *The Foundations of the Science of War*, 107-10.

50. Charles Andrew Willoughby, *Maneuver in War* (Harrisburg, PA, 1939), 17. Willoughby called Fuller "the father of the principles of war" and "an Anglo-Saxon Clausewitz." See ibid., 28, 30.

51. Colonel L. D. Holder, "Educating and Training for Theatre Warfare," *Military Review* 9 (September 1990), 88.

52. Franz, "Grand Tactics," 33. To an American in the 1920s the German term "operations" would most likely have meant "strategy," presumably military strategy, while the German term "strategy" equated to "grand strategy." James S. Corum, *The Roots of Blitzkrieg: Hans von Seeckt and German Military Reform* (Lawrence, KS, 1992), xiv.

53. Karl-Heinz Frieser, "The Execution of the 'Sickle Cut Plan' Using the Example of Panzergruppe [Armored Group] Kleist," in *Operational Thought*, 84. Frieser also notes that the Second Battle of Sedan, which saw 1.5 million Allied soldiers trapped in a 400-kilometer-long envelopment, represented a substantial leap in spatial terms when compared to the 120,000 Frenchmen surrounded in a nine-mile circle at Moltke's 1870 "Cannae of the Nineteenth Century." See ibid., 99.

54. On one route the *Panzergruppe* would have stretched 1,540 kilometers, the distance from the Luxembourg border to beyond Koenigsberg in East Prussia. See ibid., 87. In the 1990-1991 Gulf War over 65,000 armoured and support vehicles were moved to an attack position on Iraq's right flank. *Conduct of the Persian Gulf Conflict: An Interim Report to Congress*, July 1991, 4-6, 4-23.

55. Frieser, "The Execution of the 'Sickle Cut Plan,'" 100.

56. David M. Glantz, "The Nature of Soviet Operational Art," *Parameters* 1 (Spring 1985), 2-7; and David Glantz, "Soviet Operational Formation for Battle: A Perspective," *Military Review* 2 (February 1983), 3-11.

57. Colonel David M. Glantz, *Soviet Tactics at Kursk, July 1943* (Fort Leavenworth, KS, 1986), 1, 24-25, 40-42, 61-66. Kursk remains a classic example of the offensive-defensive.

58. On the advantages and disadvantages of encirclements, see General S. M. Shtemenko, *The Soviet General Staff at War 1941-1945*, vol. 2, translated by Robert Daglish (Moscow, 1981), 311-14.

59. Gerd Niepold, *Battle for White Russia: The Destruction of Army Group Centre June 1944*, translated by Richard Simpkin (London, 1987), 262. Here Soviet planners adhered to the operational principle "Think ahead, don't order ahead," while logically

developing "centres of effort" to deliver a "main punch."

60. Christopher Duffy, *Red Storm on the Reich* (New York, 1991), 10-14, 29-113, 313-14.

61. Sixty-two percent of Soviet tanks in Operation Bagration were allocated to mobile groups. See Niepold, *Battle for White Russia*, 263.

62. Charles Dick, *The Operational Employment of Soviet Armour in the Great Patriotic War* (Sandhurst, 1988), 1-4, 27, 46-64.

63. Glantz, "The Nature of Soviet Operational Art," 10-11.

64. Dick, *The Operational Employment of Soviet Armour*, 2.

65. Troop density in Normandy was estimated in 1944 to be 2.5 times that of the Eastern Front. See English, *Normandy*, 231.

66. John Keegan, *Six Armies in Normandy* (New York, 1982), 314-16. In comparison, Stalingrad cost the Germans 20 divisions.

67. The Allies aimed to reach the Seine by D+90 after D-Day, but they established themselves along the river on D+80. Carlo D'Este, *Decision in Normandy* (London, 1983), 432-33.

68. Ladislas Farago, *Patton: Ordeal and Triumph* (New York, 1975), 546-47. Patton had, of course, studied the conduct of war in great depth and could be considered "self-taught" in the operational art. See Steve E. Dietrich, "The Professional Reading of General George S. Patton, Jr.," *The Journal of Military History* 4 (October 1989), 387-418. That Montgomery fought largely to retain the "initiative" in Normandy as opposed to merely seizing "ground" also lends his operational approach a degree of sophistication not always recognized.

69. "Comments on Recent Operations (Lessons Learned)," 18 March 1945," Simpson Papers, U.S. Army Military History Institute, Carlisle Barracks, PA.

70. Almost 80 percent of the 1,213 warships that participated in Operation Neptune during the D-Day invasion of Normandy were British and Canadian. Only slightly more than 16 per cent were American and another 4.5 percent flew French, Dutch, and other Allied flags. David G. Chandler and James Lawton Collins, Jr., eds., *The D-Day Encyclopedia* (New York, 1994), 380. Operation Husky, the Allied invasion of Sicily in July 1943, was the largest amphibious operation of the modern era.

71. Significantly, in that war Israeli doctrine also included the tactical, operational, and strategic levels of war. See Franz, "Grand Tactics," 34; and General Donn A. Starry, "A Tactical Evolution — FM 100-5," *Military Review* 8 (August 1978), 2-11.

72. Which was "sometimes also known as military strategy," according to Colonel Wallace P. Franz. "This level," he added, "comprises the art of war as opposed to the art of fighting (tactics) and has been avoided like the plague in the U.S. Army." Ibid., 32. A similar view had been offered in Edward N. Luttwak, "The Operational Level of War," *International Security* 3 (Winter 1980/81), 61-79. See also Lieutenant Colonel Scott A. Marcy, "Operational Art: Getting Started," *Military Review* 9 (September 1990), 106-12.

73. The principles of war were also included in the 1982 and 1986 versions of FM 100-5.

74. Colonel Harry G. Summers, *On Strategy II: A Critical Analysis of the Gulf War* (New York, 1992), 134-35, 139-49; Romie L. Brownlee and William J. Mullen III, eds., *Changing an Army. An Oral History of General William E. Dupuy* (Carlisle, PA and Washington, DC, 1988), 188-92; and David Jablonsky, "US Military Doctrine and the Revolution in Military Affairs," *Parameters* 3 (Autumn 1994), 21-23.

75. William S. Lind, *Maneuver Warfare Handbook* (Boulder, CO, 1985), 4-24, 30-31, 69-70, 73-74. See also Richard D. Hooker, Jr., "The Mythology Surrounding Maneuver Warfare," *Parameters* 1 (Spring 1993), 27-38, and "Attritionists — or Technologists — vs. Maneuverists" 2 (Summer 1993), 107-10. See also Lieutenant Colonel Paul T. Devries, "Manoeuvre and the Operational Level of War," *Military Review* 2 (February 1983), 13-34.

76. British Army doctrine described "Operational Art" as the "skilful execution of the operational level of command," defined as the third of the following four levels: *grand strategy*, "the application of national resources to achieve policy objectives"; *military strategy*, "the application of military resources to achieve grand strategic objectives"; *operational*, "concerned with the direction of military resources to achieve the objectives of military strategy"; and *tactical,* "the direction of military resources to achieve operational objectives." *Design for Operations — The British Military Doctrine*, Army Code No. 71451 prepared under the direction of the chief of the General Staff, 1989, 37-39.

77. Brigadier Richard Simpkin, copy of "Introduction on Definition of 'Operational,'" December 1983, 1-3 (provided to author).

78. Richard E. Simpkin, "Manoeuvre Theory and the Small Army," *British Army Review* 78 (December 1984), 5-6, and his *Race to the Swift* (London, 1986), 23-44. In 1992 a Canadian officer argued that the operational art could be exercised in "lesser operations such as ensuring the safety of national citizens in a foreign country or providing aid to the civil power." Colonel K. T. Eddy, "The Canadian Forces and the Operational Level of War," *Canadian Defence Quarterly* 5 (April 1992), 21.

79. Simpkin, "Manoeuvre Theory," 12-13.

80. See also Major C. S. Oliviero, "Manoeuvre Warfare: Smaller Can Be Better," *Canadian Defence Quarterly* 2 (Autumn 1988), 67-72.

81. See, for example, Ian Brown, "Not Glamorous, But Effective: The Canadian Corps and the Set-piece Attack," *The Journal of Military History* 3 (July 1994), 421-44.

82. Simpkin, "Manoeuvre Theory," 13; and *Race to the Swift*, 22-23.

83. Simpkin, *Race to the Swift*, 297-321; and Franz Uhle-Wettler, *Battlefield Central Europe: Danger of Overreliance on Technology by the Armed Forces* (manuscript copy, undated), 3, 56-70.

84. David R. Jones, "Ogarkov and the Evolution of Operational Art," *Tactics and Technology* (Toronto, 1986), 15-16; C. J. Dick, "Soviet Doctrine, Equipment Design and Organization," *International Defence Review* 12 (1983), 1715-16; Richard Simpkin, *Red Armour* (London, 1984), 158-59. See also Simpkin's *Deep Battle: The Brainchild of Marshal Tuchachevskii* (London, 1987) and John Erickson, Lynn Hansen, and William Schneider, *Soviet Ground Forces: An Operational Assessment* (Boulder, CO, 1986).

85. Roy Allison, "New Thinking About Defence in the Soviet Union," in Ken Booth, ed., *New Thinking About Strategy and International Security*, (London, 1991), 226-28, 231; and Raymond L. Garthoff, *Deterrence and the Revolution in Soviet Military Doctrine* (Washington, DC, 1990), 84-85.

86. Christopher Donnelly, "The Development of the Soviet Concept of Echeloning," *NATO Review* 6 (December 1984), 9-17; and P. H. Vigor, "Soviet Echeloning," *Miltary Review* 8 (August 1982), 69-74.

87. General Bernard W. Rogers, "Follow-on Forces Attack (FOFA): Myths and Realities," *NATO Review* 6 (December 1984), 1-9; and Richard Mills, "Follow-on Forces Attack: The Need for a Critical Assessment," *Defence Yearbook 1990* (London, 1990),

125-37.

88. John A. English, "Central Army Group: Pillar of NATO Defence," in Ian V. Hogg, ed., *Jane's 1982-83 Military Review* (London, 1982), 107-23.

89. Glantz, "The Nature of Soviet Operational Art," 2.

90. John Gooch, "Soldiers, Strategy and War Aims in Britain 1914-18," in Barry Hunt and Adrian Preston, eds., *War Aims and Strategic Policy in the Great War* (London, 1977), 37.

91. See Clausewitz, *On War*, 357-59, 379-80, 488.

92. David Jablonsky, "US Military Doctrine and the Revolution in Military Affairs," *Parameters* 3 (Autumn 1994), 23-27.

93. Lieutenant General Sir Garry Johnson, "On Winning: An Option for Change without Decay," *The RUSI Journal* (Autumn 1991), 13.

94. E. M. Forster, *A Passage to India* (London, 1989), 190.

3

The Operational Art: The Elder Moltke's Campaign Plan for the Franco-Prussian War[1]

Bradley J. Meyer

There have only been two occasions in modern European history when a single great power has defeated another great power. Those two occasions were the defeat of Austria in 1866 by Prussia, and the defeat of France in 1870-1871 by a Prussian-led alliance,[2] whose members all became part of modern Germany in 1871. On both of these occasions, the same general directed the Prussian campaigns: Helmuth von Moltke, chief of the Prussian General Staff from 1858 to 1888. The leverage offered by what one would today call operational art was an important element in the speed and completeness of these Prussian victories.

Operational art as practiced by von Moltke had three main elements. The first was a clearly defined goal or objective for the campaign. In planning for the Franco-Prussian War, Moltke chose the early defeat of the French field army as the objective of the Prussian campaign. The second element of Moltke's operational art was a selection of goals to which his operations would be directed. Three operational goals influenced his campaign planning: to bring the numerical superiority of the Prussian army to bear, to exploit a gap which existed between the initial French concentrations at Metz and Strasbourg, and to attack the enemy simultaneously in the front and the flank. The third element was an understanding of the mechanics of operations: for instance, how to move large bodies of troops efficiently along roads and deploy them in fighting formation. Manipulation of the mechanics of operations helped Moltke to achieve his operational goals and, ultimately, his campaign objectives.

A word about operational goals is necessary. The operational artist makes decisions on the basis of his objectives, not on that of strategy and tactics. Although important determinants in establishing operational goals, strategy is too broad and tactics too detailed to help the operational commander make decisions in the light of battlefield circumstances.

Command and control in war can be thought of as a cybernetic control loop. Cybernetic control loops are simple—they involve a desired state, a means of observing the environment, and a way of affecting the environment.[3] Moltke used his operational goals in an analogous fashion: as a goal or desired state toward which he directed operations, adjusting his operational decision making to circumstances as he went along. Of course, Moltke would not have expressed matters in such terms. However, his thinking along similar lines is undeniable. As he explained: "Certainly the field commander will keep his great goals constantly in view, undisturbed by changing circumstances, but the way in which he hopes to attain them can never be determined with certainty far in advance."[4]

THE CAMPAIGN GOAL

The most critical decision to be made in planning any campaign is the selection of its goal. If the campaign goal does not achieve the political objective for which the war is being fought, or at least contribute toward that end, then the campaign will be a waste of effort. In 1870, Moltke identified his campaign goal as the defeat of the French army:

> The operation against France will simply consist in advancing a few marches forward onto French territory, as well closed up as possible, until we meet the French army, in order to deliver battle. The direction of this forward march, in general, is Paris, since in that direction we are most likely to meet the goal of this advance, the enemy army.[5]

He believed that the military goal of the defeat of the French army would lead to the attainment of Prussian political goals.

> It is probable that a change of dynasty will occur in France after the first lost battle, and since we want nothing from France [an interesting comment in light of the eventual annexation of Alsace-Lorraine], a rapid settlement can be reached with the new power-holders in France.[6]

Opening the war by invading France and defeating the French army also fit into a larger strategic framework. Both France and Austria were hostile to Prussia and, taken together, their armies were stronger than those of Prussia. "If we turn half of our army against each of these two powers," Moltke noted, "then we are superior [in numbers] to neither of them."[7] The only question for Moltke was against which of these potential enemies Prussia should defend with weak forces in order to advance against the other "as strongly and as offensively as possible."[8] If Prussia chose to invade Austria, the Austrians, who needed more time to mobilize than did the French, could avoid a decisive battle and retreat into the depths of their country. Moltke calculated that a Prussian offensive against Austria might come to a halt in front of the Austrian fortress of Olmuetz, not far from Vienna, or on the Danube. Meanwhile, the French might achieve success in their expected invasion of southern Germany.[9] On the other hand, if the Prussians invaded France and achieved initial success against the French army, the Austrians might decide not to enter the war at all, and "let the half-drawn sword fall back into the scabbard."[10] That is exactly what happened. In the face of French setbacks in the opening battles of the war, the Austrians did not enter the war against Prussia. The Franco-Prussian War was thus a duel between France and Prussia. The success of Moltke's campaign goal—an early defeat of the French army—and the success of Bismarck's diplomacy enabled Prussia to avoid the specter of a two-front war, which haunted united Germany in the First and Second World Wars.

Moltke's decision to lead the initial Prussian advance toward Paris—the political and cultural center of France—was designed to ensure an early battle victory. "The direction of this forward march, in general, is Paris, since in that direction we are most likely to find the goal of this advance, the enemy army."[11] Paris was a strategic objective of such importance that the French would be forced to defend it. The importance of Paris and its location relatively close to the border with Germany put the French on the horns of a dilemma. If the war were viewed as a purely military contest, it would make sense to fall back into the interior of France, which would stretch Prussian supply lines and provide time for integrating the reserve forces into the army. France had a larger navy than Prussia, so the western ports of France could never be cut off by sea. But no government could contemplate abandoning Paris. Apart from its emotional significance, it was unlikely that any government could govern France without Paris.[12] By aiming his advance at the strategic objective of Paris, Moltke could be sure of engaging his real target, the French army.

OVERVIEW OF THE CAMPAIGN

What was Moltke's mental picture of the campaign plan for the invasion of France? The assembly area—the place where divisions, corps and armies were assembled into organisms that could march and fight—was the Pfalz, in the Rhineland. Selecting the assembly area was one of the most important decisions Moltke made in planning the campaign.[13] The main reason he selected the Pfalz was that it was a good jumping-off place for his intended invasion, although other reasons will be addressed below. The constituent units of the divisions, corps, and armies—namely, regiments, batteries, supply and equipment trains, and supporting units—arrived in the assembly area by means of the *Aufmarsch* or "approach march." It mostly occurred by rail, but all or part of it might occur by foot. Once the units were assembled, the army moved forward—the *Vormarsch* or "forward march"—ready to give battle.[14] Of course, the advance guard would have to hold up any approaching enemy long enough for the force to deploy. The advance was made in an "assembled" fashion, a term first used by Napoleon and meaning that all the major units could march to each other's assistance and deploy for battle within a day's time.[15]

According to Moltke's campaign plan, the Prussian army assembled in the Pfalz. The main mass, composed of the Second and Reserve Armies, 150,000 men, then moved forward in three main columns, two echelons deep. The front of the advance was two marches wide and two marches deep (a "march" was basically the distance a column could cover in a day's time). This formation is an example of the third major division of operational art postulated in the introduction to this chapter: controlling and manipulating the mechanics of operations. The point of the formation was that the whole mass of the Prussian center armies, the Second and the Reserve, could deploy for battle either to their front or to either flank.[16] The Prussian First Army, 60,000 strong, covered the right flank while the Third Army, 130,000 strong, covered the left.[17] In addition, the First Army could potentially serve as an "offensive flank,"[18] able to swing around and hit a French force in the flank while the Prussian center armies attacked the French frontally. One of Moltke's operational goals was to produce such a situation—to attack the enemy simultaneously in the front and flank. He described the planned forward march in the spring of 1870:

> On the direct way from the Pfalz to the French capital lies Metz [a major fortress]. This place will be gone around to the left and need only be observed. The next strategic approach march, if there has not already been a battle, is the line on the Meuse between Lunéville-Pont à Mousson. . . . On the line Lunéville-Pont à Mousson we have two railroads to the rear; there, if not earlier, there must probably come a decision, and beyond that nothing can be foreseen in detail.[19]

At that point, campaign planning for the war came to an end. Moltke expected a battle shortly after he crossed the French border. He planned to swing to the left of the fortress of Metz and to approach the Moselle between Lunéville and Pont à Mousson; but the details of those movements would be decided in light of the situation when the Prussian armies arrived in that area, when information concerning the whereabouts of the French army would presumably be available. At that time operations would be ordered so as to bring about battle with French forces under conditions as advantageous to the Prussians as possible. After crossing the French border, the control of operations would be a matter for day-to-day decision-making within the context of an ongoing campaign, not a matter for campaign planning.

There was a detailed timetable for the forward march from the assembly areas. It listed each of the army corps and what towns they would reach by the end of each day.[20] Somewhat unusually, since Moltke normally operated by means of what today is called "mission orders," he specified that the marches of all divisions and corps be "specially regulated from above."[21] He did not believe it possible to plan a campaign from beginning to end, as one might write the script of a play:

> It is a fallacy, when one believes it is possible to determine a plan of campaign far in advance and carry it out to the end. The first clash with the enemy main force creates, according to its result, a new situation. Many things cannot be carried out which one may have intended, many things become possible which were previously not expected. To understand the changed circumstances, on that basis to direct what is suitable and carry it out in a determined fashion, is all that the army leadership can do.[22]

OPERATIONAL GOALS

Looking at the Prussian army's operational goals is useful both in determining

Moltke's choice of his opening move—the assembly in the Pfalz and the forward march to the French border—and in determining his actions after the "first contact with the enemy." At that point, he had to make operational decisions in the course of an ongoing campaign. For Moltke, this meant primarily the direction in which he should order his armies to move, based on his mental picture of how the campaign was unfolding and of how he wanted it to do so. For instance, some of Moltke's forces might move past Metz on the way to Paris; if information became available to the effect that the French still had major forces in Metz, then Moltke might want to turn some of his units to the north and cut off any French retreat from Metz, even if this required a major battle. This is exactly what happened, and such movements were ordered, leading to the battles of Mars-La-Tours and Gravelotte St. Privat, and the shutting up of Bazaine's army in Metz.[23]

Deploying from a marching column to a fighting formation was another operation of war. The timing of this was important: if begun too soon, the army would have too far to move forward in fighting formations; if begun too late, the enemy might disrupt the whole operation. Such operations would likely be under the control of individual army commanders, as Moltke was concerned only with coordinating the cooperation of different armies in a given battle. Then again, if Moltke did not want a battle to occur at a given time and place, he would seek to control operations to that end.

Operational goals were a better guide for controlling operations than was the campaign objective, the defeat of the French army. In a sense, operational goals were intermediate objectives, leading to the campaign objective but closer to the level of individual operations and, therefore, a better guide to operational decision-making. The first, and most basic, of Moltke's operational goals was to seek out the enemy's army and bring it to battle. He had a superiority in the number of troops available, and he meant to bring it to bear. There were strategic reasons for wanting a quick victory over the French, namely that it would discourage other powers who might want to enter the war. Moltke noted that "the operations plan for the offensive against France . . . consists only in seeking out the main force of the enemy and, where one finds it, to attack. The difficulty lies only in carrying out this simple plan with very great masses [of troops]."[24] This last sentence is the closest Moltke comes to humor in his campaign planning. He understood that he would be attempting to control armies of a scale that had only been encountered in the 1813 campaign in Germany, when the size of armies had overwhelmed the ability of both the great Napoleon and the allied powers to control them effectively.

In planning for the war with France, Moltke calculated that he would have at least 360,000 troops for the invasion, more if Bavaria participated. The French would have around 250,000.[25] The intended Prussian forward march from the assembly area to the Meuse was about 50 English miles. In Moltke's eyes, the only difficulty connected with this march, apart from the anticipated French resistance, was the narrow space in which the movement of armies totaling nearly 400,000 men had to be carried out.[26] Basically, the Prussian armies were aiming for the gap between Metz and the Vosges Mountains, a distance of about 60 English miles.[27] Since the Pfalz is a hilly region of Germany, the road net was not particularly dense.

So long as the French had to get from their marching formation to a fighting formation, the difficulties would apply equally to both sides. The possibility existed, however, that the French would deploy for battle and wait for the Prussians to stumble onto them. It was even possible that they would deploy for battle and attempt to move forward in battle formation, cross-country, without shaking out into columns marching on the roads. This could be done. It was slow, but the French might be able to hit a Prussian column before it could fully deploy for battle. Moltke's solution to these potential difficulties was to make sure that the advance guard was of sufficient strength, and far enough forward, to deploy fully for battle. He exhorted his subordinates to keep lateral communications constantly in mind as the main army moved forward, so that it could deploy to either flank as well as forward.[28]

Another of Moltke's operational goals was to drive between the French concentrations that were likely to form at Strasbourg and Metz, and operate on the inner line between them. He noted that if the French wished to mass forces close to the German frontier quickly in the earliest phase of the war, they would almost inevitably do so in both Metz and Strasbourg. He drew this conclusion after studying French railway lines and their capacity. Both Metz and Strasbourg were major rail junctions. It would be easy for the French to mass troops at those two points. On the other hand, once there, these forces would be separated by about 80 miles and the Vosges Mountains.[29] The French commanders would be able to unite them only by means of foot marches. By assembling in the Pfalz, the Prussians would stand on the inner line of operations between the two French forces and could turn against one or the other, or both, if, as Moltke put it, "we are strong enough."[30] At least initially, Moltke believed that the two French forces would be "without the possibility of mutual support."[31] At that time, two forces were said to be out of supporting distance if they were more than a day's march from each other.

It was likely that the French would attempt an early concentration close to the border. Their military system of a standing army of long-service professionals gave them the ability to jump off quickly and, perhaps, disrupt the assembly of enemy forces or launch a preemptive offensive of their own.[32] This was an advantage over the Prussian system, heavily dependent on the mobilization of reserves. Six main rail lines ran from the interior of Prussia to the Pfalz, which would allow for the rapid assembly of forces. From the Pfalz, Moltke planned to move forward so as to drive between the French forces at Metz and Strasbourg.[33] At the same time, he wanted to be able to attack either force simultaneously in the front and flank,[34] while the other French force was inactive or being delayed by Prussian action. Moltke thus planned to operate on "interior/exterior" lines, whereas in the war with Austria he had operated on purely exterior lines. The choice between exterior and interior lines was made on purely practical grounds, in this case the probable existence of a gap between the French forces at Metz and Strasbourg.

Driving between the French Metz and Strasbourg forces paid off for the Prussians. The Third Army engaged the French Strasbourg force at the Battle of Froeschwiller. Tactically, the battle was a draw. The French Chassepot rifle proved quite capable of repelling unsupported Prussian infantry attacks, but, gradually, Prussian artillery began to take effect and Prussian units appeared on the French flanks. The most important result of this battle was that two French corps, in their retreat, took a road that led them away from, rather than toward, the larger French Metz force. This allowed Moltke to concentrate on the French Metz force of five army corps, defeat it, and cut off its line of retreat to the northwest. That force then retreated into Metz, where, after being besieged, it eventually surrendered. The French Strasbourg force, meanwhile, retreated to Chalons. Augmented by a corps, which had not been engaged in the opening battles, and a newly formed corps, this force attempted to relieve the siege of Metz. But it was driven north toward the Belgian border and destroyed in the Battle of Sedan.[35]

By having clear operational goals in mind, observing the relative position of the contending forces—as clearly as he could discern it through the fog of war—and using operations to improve his situation, Moltke was able to convert small advantages into large advantages. A case in point is the Strasbourg force's divergent line of retreat after Froeschwiller. He was able to force the French into a descending spiral,[36] a progressively worsening situation in terms of the framework of the campaign which he had succeeded in creating.

The opening battles at Spicheren and Froeschwiller on 6 August 1870, although they were French defeats, were not decisive. The Metz force and the Strasbourg force did not establish contact with each other during their retreats, and the Metz force retreated to the temporary safety of the fortress of Metz. Once it became clear that the French would stay at Metz, Moltke's efforts were guided by another operational goal—to drive French forces to the northwest, away from Paris, and to force them against the Belgian border. This objective found expression in written orders on 16 August, ten days after the opening battles of the war.[37] It would be advantageous for the Prussians if the French armies were cut off, not only from Paris, but from a retreat into southern France, where there were rich supplies of food, men not already called into military service, access to arms and supplies, and where, conversely, the Prussians would find it difficult to pursue and fight due to lengthening supply lines.[38]

The result of this operational goal, and the operations that resulted—the Battles of Mars-La-Tour and Gravelotte-St. Privat—saw the Metz force besieged in the fortress of Metz. The operational goal of driving French forces away from Paris to the northwest also found expression later in the campaign. This occurred when a French force, foolishly sent to relieve Metz by clandestinely moving between the Belgian border and the Prussian armies advancing on Paris, was driven near the Belgian border, encircled, and destroyed in the Battle of Sedan. This battle effectively destroyed those elements of the French regular army not already encircled in Metz.[39]

The operational goals discussed so far arose from the particular circumstances of the Franco-Prussian War. They were not of general applicability in a war fought by another power in the same era, or even in a war fought between Prussia and, say, Austria. Another operational goal, that of arranging operations so as to attack the enemy simultaneously in the front and the flank, was of general applicability for that era and time. It was in fact this goal that had brought Moltke great success in the war against Austria in 1866, culminating in the Battle of Koeniggraetz.

The idea of attacking the enemy simultaneously in the front and flank grew out of an analysis by Moltke of the possibilities of military operations. The analytical process he utilized was not "finger-tip feeling" (*Fingerspitzengefuehl*) but, rather, a logical and clearly explained analysis of the goals of military operations aimed at defeating an enemy. He wrote a clear exposition of the rationale for his operational goals in 1869, "The Instructions for the Higher Troop Commanders," which the king of Prussia approved.[40] Moltke began by discussing the increased effect of infantry firepower due to constant improvements in the basic infantry weapon, the rifle. He observed that, by 1869, a slight fold of ground with a long, clear field of fire in front would enable the defense to beat off the heaviest and most determined attack. Barriers—streams, ditches, walls—were no longer needed to make a position unassailable; the full development of firepower was crucial.[41] As he put it,

> One can hardly question, that the man who stands and fires has an advantage over the man who fires while moving forward, that the one finds in the terrain a support, the other a hindrance, and that when the most eager elan comes up against a calm steadfastness, then the fire effect, which in our time has risen so exceedingly, decides.[42]

Moltke drew the obvious conclusion that at the tactical level, at least, it was easier to defend than to attack. "Every victorious battle must . . . end with an offensive advance; one can question, however, whether it should therefore begin with an offensive advance."[43] Since the planned campaign against France was an offensive one, Moltke faced the problem of how to wage an offensive campaign in the face of the power of the tactical defensive.

If frontal attacks had become difficult, if not impossible, perhaps it would be possible to go around a flank. But Moltke foresaw difficulties with that procedure if the outflanking force began operations from in front of the enemy position. He noted that the effective range of infantry and artillery had increased, making it necessary for the outflanking force to swing wider around the defensive position. Then, again, the defending forces were likely to echelon their reserves more deeply—to put them further to the rear—in order to get the reserves beyond the increased range of the attacker's fire. This meant that a flank attack would have to swing farther around a defensive position, far enough to get behind the deeply echeloned reserves.[44]

At this point, the mechanics of operations became important. Once troops got as close as the roads would take them to a good place to form up for a flank attack, they would have to get off the road, perhaps moving some distance to form up for the flank attack; only then could they deploy into line for attack. The troops that moved into line first would have to wait until the last troops deployed. Then the whole mass could move forward confidently.

As Moltke pointed out, with armies as large as they were, an effective flanking force would also have to be fairly large. The larger they were, the more subject to delays they were in deploying. If the flank attack had to swing wider, due to the increased range of artillery and the associated deep echelonment of the defender's reserves, then the whole evolution would likely take more than a day's march. Specifically, Moltke calculated that the flanking force would have to travel 15 miles to be effective if launched from a position in front of the enemy. With delays in getting under way and deploying for battle, it could not complete the whole evolution in a day's time. Meanwhile the enemy might be doing any number of things: shifting forces to meet the flank attack, attacking the force left behind to hold its immediate front, or withdrawing from the field altogether.

Moltke's solution was to launch the flank attack from what today is called an operational distance:

> Matters go incomparably better when on the day of battle the [friendly] forces can be concentrated from separated points against the battlefield itself, when in other words the operations [*die Operationen*] have been so conducted, that from different sides a last short march leads simultaneously against the front and flank of the enemy. Then strategy [*Strategie*] has achieved the best which it is able to achieve, and great results must be the consequence.[45]

Moltke had pursued this operational goal with great effect against Austria in 1866. In that war, the flank attack of the Prussian First Army, which sealed the fate of the Austrian army at Koeniggraetz, had been launched from the Prussian border. The Prussian First and Second Armies did not come into contact with the Austrians until the day of the climactic battle itself. When, in the course of the campaign, the two Prussian forces threatened to come too close together, to unite before the decisive battle, Moltke deliberately kept them separated so as to be able to catch the Austrian army between them.[46]

The operational goal of attacking the enemy simultaneously in the front and the flank arose from factors that were operating in Moltke's day, specifically the interaction between the size of armies, the power of defensive weapons, and the mechanics of marching troops. Moltke's operational goals were different from those of Napoleon, just as Schlieffen's operational goals differed from those of Moltke. Operational goals, and therefore operational art, arise from the underlying nature of warfare itself, and the factors which affect warfare. It follows that the operational art changes over time because warfare itself changes.

Apart from the objective of the 1870 campaign itself, which was the defeat of the French army, the most important campaign-planning decision made by Moltke was the choice of the location of the initial assembly of the army—the Pfalz. Moltke himself said, "A mistake in the original assembly of the armies can hardly be put right again in the whole course of the campaign."[47] If a mistake was made in this critical "opening position" of the game, it would be difficult to "catch up" as the enemy maneuvered to attempt to avoid being put at a disadvantage—moreover, the enemy would already be a move ahead.

Moltke chose to assemble his armies in the Pfalz because it was a good jumping-off point for his plan to divide French forces Metz and Strasbourg. But he also considered the risks of a deployment in the Pfalz. First, there was the matter of whether the French could interfere with the assembly of his forces. Moltke's plan called for the forces of the North German Confederation to first mobilize at their garrison station where reservists would be called in and war equipment issued. They would then be transported, mostly by rail, to their assembly areas in the Pfalz. By the twenty-fourth day of mobilization, Moltke calculated, all the assembled units would be supplied with wagon trains, and the whole mass could move forward against the French. The French, however, had a large professional standing army stationed in northeastern France. If they chose to forgo the process of mobilization—that is, absorbing and equipping their reserve troops—they could form field armies from the peacetime cadres of their units. At least theoretically, the French might then attempt to disrupt the Prussian movement to the frontier.

Moltke satisfied himself that they could not. Advance guards near the border would delay the attacking French or, at least, give information as to their whereabouts, in which case the unloading of the trains carrying the bulk of the Prussian army could be moved back toward the Rhine or beyond. Moltke calculated that if this movement was necessary, by the fourteenth day after mobilization, he could oppose the French with equal force along the Rhine; a few days later, with the Rhine fortresses acting as fortified bridgeheads, he could give battle with a numerical superiority over the French of two to one. However, he judged that the French would not attempt this particular ploy.[48]

This case covered a French frontal attack against the assembly of the main Prussian forces. Another possibility was that the French would attempt to go south of the main Prussian assembly and invade the South German states. Moltke calculated that he could defend against this thrust by means of a flank attack from the Pfalz. Similarly, if the French went through Belgium to attempt a crossing of the lower Rhine, they could also be defeated by means of a flank attack.[49]

The use of such "flanking positions" was one of Moltke's favorite operational solutions for dealing with defensive problems. If his defenders took up a position to one side of the route that an attacking force would take toward its objective—the "flanking position"—the attackers would have just two choices. They could turn aside to deal with the flanking force. Regardless of the outcome of any battles that might result, this would temporarily delay their assault. Or, the attacking force could press on toward the objective. In the latter case, the defenders could move forward and cut the lines of communication of the attacking force. The attacking force would then face battle without either a supply line or a secure line of retreat.[50] There was therefore no reason for Moltke to change his assembly area to meet possible French offensives through Belgium or into southern Germany; the assembly area in the Pfalz could serve as a flanking position to counter either of these possibilities, as well as being a good jumping-off position for the intended attack into France.

Similarly, Moltke calculated that the three army corps he planned to leave in eastern Germany could counter any possible operations mounted against Berlin by Austria. The three eastern army corps could accomplish this task mainly by means of flanking positions, strengthened by rivers and fortresses in the area.[51]

Moltke's campaign planning did not go beyond the choice of his initial assembly area. As he once noted, "Only the layman expects to see in the course of a campaign the consistent carrying out of an original design, conceived in advance, considered in all details, and adhered to the end."[52] But, as he also observed, "A mistake in the original assembly of the armies can hardly be put right again in the whole course of the campaign."[53] He knew that just as the opening move of a chess game often determines its character, the initial deployment in a military campaign would tend to set the campaign running along certain channels. In the case of his planned invasion of France, Moltke wanted to enter the initial battles under advantageous conditions. That meant attacking the enemy simultaneously in the front and flank, defeating the Strasbourg and Metz forces, and driving the French forces northwest, away from Paris. The initial deployment area in the Pfalz was a good starting point from which to move on to Paris, the ultimate military and political objective.

Once the 1870 campaign was under way, the relative positions of the two opposing forces changed as they maneuvered against each other. Battles destroyed all or part of the contending forces, forced retreats, and demoralized or encouraged one side or the other. But Moltke's goals remained the same. He tried to achieve those goals by sustaining advantageous conditions of engagement within the framework of the mechanics of operations—what his forces could achieve in terms of marching and fighting.

THE MECHANICS OF OPERATIONS

A single Prussian army corps, with all its infantry, cavalry, artillery, and essential support elements, took up slightly less than 13 miles of roadway.[54] Since a normal day's march for a corps was about 10 miles, it followed that the head of the column would arrive at its destination before the tail had set out. Although it might take an individual soldier only four or five hours to make the day's march, it took about six hours for the tail of the column to close up with the head; thus, it would take the corps as a whole at least a day to complete a march. Furthermore, under normal circumstances, no more than one corps could be moved down a given road if there was any expectation that it would arrive at a given point on the same day.[55]

On a long route march, with no prospects of contact with the enemy, it would be unnecessary to close up the tail of a column with the head each day. Units would simply fall out to the side of the road when they stopped marching and start again the next day. But, if a battle were to occur, it would be necessary for the tail to close up with the head in order for the full combat power of the corps to get into action on the same day. For instance, the marching corps might encounter an enemy who had occupied a position suitable for defense. An army deployed for battle could maneuver cross-country. The encounter between an army moving forward in line and the head of a column was likely to be unfortunate for the column, however desperately the column's leaders tried to deploy off the road into line of battle. Then, again, a force might deliberately deploy into battle in order to move cross-country against an undeployed or deployed column, thus avoiding the time-consuming process of marching up a column of troops and deploying them into line of battle. In planning his army's forward march from the French border to the Moselle River, Moltke worried about the French trying either of these options.[56]

Ten miles is not a long march. Sometimes the troops at the head of a corps column might march 15, 20, or even 30 miles in a day. But if they did so, these troops might outrun the ability of the column to close up behind them on the same day. Marches that might culminate in a battle took place at a slower rate than one might expect. Since it took the tail of a Prussian corps six hours to catch up with the head, it followed that a corps march in the vicinity of an enemy ought to take place under the protection of an advance guard, one strong enough, or moving far enough ahead, to guarantee the six hours needed for the tail of the corps to march up.[57]

Arrangements for the forward march of the Prussian Armies took these realities into account. The relationships described above were pushed to their practical limits in an attempt to confront the French with the largest possible force that could get into battle in a single day. The Prussian Second and Reserve Armies, 150,000 men strong, were to move in three columns and two echelons over the French border. Since there were six army corps in the Second and Reserve armies combined, each column would have two corps marching along the same road. Although this maneuver was impractical if both corps were to get into battle on the same day, a couple of methods could be employed to make it possible. The first corps would move off at first light to ensure maximum use of available daylight. They would be followed by their supply trains, which at the first sign of contact, would move off the road, freeing it for the combat echelons hurrying forward. Moltke deprecated the idea of leaving the wagon trains behind in order to accelerate marches. Anything that was not necessary for battle would be left farther behind, but most of the trains consisted of things that were essential in battle like artillery ammunition and medical units. The cavalry would already be forward, with the advance guard, thereby shortening the "march depths" by the length of road space the cavalry might occupy. Yet, the length of march had to be limited to about eight and one-half miles each day.

In this fashion, a force could be put together to concentrate 150,000 men on any given day for deployment against the enemy's front or either of its flanks[58]—a maneuver reminiscent of Napoleon's "Battaillon Carre." In 1870, a march table for the intended movement of the Prussian armies from their assembly areas to the Moselle River was designed; it showed the locations that each corps would reach every day expressed in terms of days after "mobilization day."[59] It is only when the extraordinary arrangements for this march are understood that the very dry humor in Moltke's comment can be appreciated:

> Less complicated than for the defense against Austria is the operational plan for the offensive against France. It consists only in seeking out the main force of the enemy and where one finds it, to attack. The difficulty lies only in carrying out this simple plan with very great masses.[60]

The 150,000 man force of the Second and Reserve Armies, thus, represented the largest force that could be moved across the French border in a way that it could concentrate for battle on any given day. But what about the rest of the invasion force? An estimated 60,000 troops would be with the Prussian First Army to the north and 99,000—130,000 if the Bavarians sent two corps—with the Third Army to the south.[61] With any luck, and properly coordinated by Moltke at Royal Headquarters, these forces could also be brought into the battle. This arrangement—a strong center with substantial strength on either wing —probably represented the upper limit for a force that expected to move and fight a battle on the same day in the limited room for maneuver available near the FrancoGerman border. If the Prussian armies had been smaller—say less than 150,000 men—Moltke could have commanded the main army and, in effect, also have been the campaign commander, since the only important campaign-level questions would have been when, where, and under what circumstances the main army should fight. But the Prussian armies had grown too large for such direction.

This is one reason why there has emerged what is today called the "operational art" as a subject of attention and study. By the 1860s, armies had grown so large as to require division into subordinate groups that would, hopefully, work together toward some common goal. The required coordination would have to be supplied by the campaign commander. It was precisely this type of coordination that was the great weakness of Prussian command and control during the Franco-Prussian War. Moltke might have a clear enough picture of what he wanted to accomplish, at least in general terms, but the army commanders often either did not understand what he wanted or were unable to accomplish it.[62]

In the Franco-Prussian War, the French elected to move their units to the border without completing their mobilization. The peacetime cadres took to the field without waiting in their depots to absorb the reserves, which were forced to catch up later. Although not without its own problems, France's speedy, but incomplete, mobilization posed a threat to the original Prussian assembly areas. Foreseeing the possibility of the French transforming their peacetime units into field armies without waiting for mobilization, Moltke had planned for an alternative assembly area back on the Rhine.[63] Still, once assembled in the Pfalz, the Prussian armies began their forward march to the French border. Just as in the originally planned advance to the Moselle, march arrangements were made in such a way that battle could commence on any given day with all available forces. Along the way, it was necessary for the Second and Reserve Armies to cross the Haardt Mountains, where there were only two roads available. Four army corps and some other units, 140,000 men in all, with associated trains, used one road; this required eight days to pass through the defile. The forward march had to be made in a fashion that would not expose the main force to surprise attack, with cover for its movement largely provided by the two flanking armies.[64]

Although it was important that Prussian forces be able to concentrate for battle if necessary, Moltke also wanted to ensure that concentration did not take place if no battle was in sight. When an army of the late nineteenth century approached a battlefield, the choices regarding which roads to use tended to narrow so that the army was probably better off moving cross-country; but doing so meant slowing down. If the army wanted to get up to marching speed again, it had to feed from the fields onto the local roads or, as was more often the case, road. In other words, the head of the column would march off down the road, and the rest would follow as quickly as it could. If there was an enemy nearby, this could be dangerous.[65]

Obviously, an army that concentrated for battle too soon lost a lot of maneuverability compared to enemy forces that had not concentrated. On the other hand, it was necessary to concentrate in order to fight, and an army which was caught on the march could be overwhelmed by a concentrated army. The key was timing: concentration should occur neither too early nor too late. Stage-managing the arrangements for the final concentration in light of circumstances at the time—the position of one's own forces, that of enemy forces, or the local road net—was an essential aspect of generalship. These considerations led Moltke to conclude as follows:

> A lasting concentration is a calamity, often an impossibility . . . it pushes toward a decision, and cannot therefore be allowed to occur, when the moment for a decision has not arrived. If, however, the unification of all fighting forces for the battle is absolutely necessary, the arrangement of separated marches with a view toward a timely concentration is the essence of strategy.[66]

These are strong words. Moltke did not utter them lightly. What Moltke described here, and elsewhere, is strategy as operational art. Indeed, Moltke used the term "strategy" in a sense close to the Greek root of the term: "generalship." Today, the term "strategy" is reserved to describe the winning of wars rather than the winning of campaigns. Where Moltke divided the continuum of warfare into strategy and tactics, it is today divided into strategy, operations, and tactics. But there can be no doubt that Moltke was a practitioner of operational art as the term is understood today.

It needs to be stated that the Prussian command system proved to be weak in translating Moltke's overall guidance into appropriate action. For instance, just prior to the Battle of Spicheren, the initial encounter with the French Metz force, the Prussian First Army crossed directly over the intended path of the Prussian Second Army—twice.[67] Nevertheless, as Michael Howard has pointed out, the initial Prussian deployment was so wide, as compared to the French concentration, that the battle was eventually decided by Prussian units coming in on the French flank. It was a matter of geometry: when the battle started, the Prussian units were more widely deployed than those of the French. By simply marching toward the sound of the guns, the Prussians were able to find the French flank.[68] But Moltke's overall direction of the campaign was responsible for the original wide deployment.

In general, during the course of the campaign, the Prussian forces moved in a more diverse and along a much longer line than did those of the French. This had three effects. First, the Prussian units moved faster because they were not clumped in a single mass that would overload the local road net. Second, the Prussians swept out over a much greater area as they advanced. This was a valuable feature in the face of often considerable doubt as to where the French main body actually was located. Largely because the Prussian forces were moving on such a broad front, the French were scooped up on their attempted retreat from Metz and on their approach to Sedan. Finally, French forces were vulnerable to flanking action, if not encirclement and destruction, when contact did occur. In comparison to the Prussian forces, the French tended to stay in clusters, moving slowly, always vulnerable to flanking attacks and encirclement. The French had forgotten the mechanics of operations that Napoleon had used so effectively in his wars only 60 years earlier. This military incompetence reached a low point when, just prior to the Battles of Mars-La-Tour and Gravelotte-St. Privat, five French army corps attempted to leave Metz by marching down a single road in file, although there were alternate routes available.[69]

Moltke devoted the rest of his life to developing a corps of officers who could understand a campaign and translate the general directives of a campaign commander into appropriate operational orders. He had 20 years to live, and he remained chief of the Prussian General Staff until three years before he died at the age of 90 in 1891. As the man who had defeated two European great powers for both Prussia and Germany, his prestige was immense. The origins of the German General Staff as a major historical actor date from the period of Moltke's leadership.[70]

CONCLUSIONS

In designing the campaign plan for an invasion of France, Moltke made two basic decisions: where to assemble his army and how to execute the forward march to first contact with the enemy. Everything else was a matter of detail, whether the arrangements for transporting the troops to the assembly area or ensuring against French or Austrian disruption of his plans. Both of these decisions were made on the basis of the campaign's objective and its operational goals.

The Prussian campaign plan for the war against France served its purpose. Despite serious errors in both the execution of Moltke's operational plans and its intentions, a course was set from which the French were unable to recover. Moltke's operational goals ultimately paid off, particularly the idea of driving a wedge between the French forces at Metz and Strasbourg. Not all of this was a function of campaign planning. Much of Prussia's success can be attributed to operational decision-making—guided, of course, by the broader campaign and operational objectives—during the course of the campaign.

Moltke's campaign planning for the 1870 war was notable both for its completeness and its recognition that not all variables are predictable. Some of the planning was very detailed—the transportation of troops to assembly areas and the forward march were both laid out in precise timetables. Concerning these arrangements, Moltke noted, "There remained in the completed arrangements nothing to change, only to carry out the previously thought out and previously prepared."[71] But he also expected that the final movements to the opening battle would be subject to adjustments based on changing circumstances. His understanding of what could and could not be planned in advance reflected his deep insight into contemporary warfare.

Moltke's operational planning and decision making depended on his operational goals. The campaign objective—"to defeat the French army"—made strategic sense, and it represented a military objective that would attain the political objectives of the war. It also represented a strategic choice—France before Austria. But this campaign objective was not a viable basis for campaign planning or for operational decision-making during the course of the campaign itself. The question would immediately arise: how should the French army be defeated? Operational goals provided the answer to this question, when combined with the peculiar circumstances of every decision and an understanding of the mechanics of operations. Saying "defeat the French army" does not provide any clues to a course of action. But saying "defeat the French army by bringing the numerical superiority of the Prussian forces to bear" provides a basis for planning, when combined with a knowledge of how to maneuver large forces.

Commanders who wish to exploit the phenomena of operational art cannot function without operational goals. Without operational goals, there is no basis for operational planning or decision-making. But devising good operational goals is difficult. In a sense, although deriving operational goals must be based on a process of reasoning, it is the most creative act of the operational commander.

The derivation of an operational goal in peacetime, such as Moltke's operational goal of attacking the enemy simultaneously in front and flank, is extremely rare. Moltke based his conclusion not on battle experience but on an analysis of the changing nature and instruments of war, like improved infantry weapons and the increased size of armies. Often at the end of a long war everyone understands what works in both tactical and operational terms; but rarely are such things successfully worked out in the abstract beforehand. When they are—and when they also are effective—the effect can be devastating, as was shown in the Austro-Prussian War of 1866 and in the Franco-Prussian War.

Clausewitz's observation—"Strategy is the use of the engagement for the purpose of the war"[72]—undoubtedly spawned the often made assertion that the operational art is the "link" between tactics and strategy. Clausewitz, like Moltke, used the term "strategy" to describe a phenomenon that would generally be described as "operational art" today. Moltke realized that tactics could affect what it meant to fight a battle under conditions of advantage, which in turn could affect the operations that would bring about a battle. His operational goal of attacking the enemy simultaneously in the front and the flank was designed to manipulate the conditions of engagement for advantage. But, to do that, Moltke had to know what was an advantageous condition of engagement. Thus Moltke can be thanked for another "link" between tactics and strategy: the realization that tactical changes can have an impact on operations.

This is not to say that the operational art is all about winning battles. Both Moltke and Clausewitz recognized that winning the campaign was more important. "Many important purposes of war," Moltke observed, "can be attained without battles, through marches, through the choice of positions, in short through operations." But he also realized that "Battle is the most important means of breaking the opposing will of the enemy."[73] If it was necessary to fight battles—when, as was often the case, the enemy opposed the attainment of the campaign objective with armed force—Moltke wanted to use operational maneuvers to fight a campaign under advantageous conditions.

A good definition of Moltkean operational art is "the use of operations for the purposes of the campaign." For Moltke, operations were the raw material of the strategist. He used the operational art to his advantage, and there was probably no one who better understood the operational art of his day. But most of what we would describe as operational art he described as strategy, as for instance when he said that "the arrangement of separated marches with a view toward a timely concentration is the essence of strategy."[74] No one today would call the arrangement of marches a strategic matter, but it does fit well the general understanding of operational art. Moltke did not believe in general principles of war that could be applied regardless of circumstances. He did make use of general principles, but they were derived from more basic circumstances, and he was careful always not to try to use an idea beyond the bounds of its applicability. By and large, he operated on the notion: "It depends on the situation."

NOTES

1. The following chapter represents the opinions of the author alone, and does not necessarily represent the position of the U.S. Department of Defense, the Department of the Navy, the U.S. Marine Corps, the U.S. Marine Corps Command and Staff College, or the School of Advanced Warfighting.

2. The German forces in the war included the North German Confederation, of which Prussia was the dominant member, and the South German states of Baden, Wuerttemberg, and Bavaria. See Michael Howard, *The Franco-Prussian War: The German Invasion of France, 1870-71* (New York, 1962), 22-23. To conform with common usage, and for simplicity's sake, the North German Confederation and its allies will be referred to as the "Prussians" throughout this chapter.

3. Norbert Wiener, *Cybernetics; or Control and Communication in the Animal and the Machine* (New York, 1948), 115.

4. Graf Helmuth Karl Bernhard von Moltke, "Aufsatz 'Ueber Strategie' vom Jahre 1871," *Moltkes Taktisch-strategische Aufsaetze aus den Jahren 1857 bis 1871. Moltkes Militaerische Werke*, II, Die Thaetigkeit als Chef des Generalstabes der Armee im Frieden, Zweiter Theil (Berlin, 1900), 292. Unless otherwise noted, all translations are by the author.

5. Graf Helmuth Karl Bernhard von Moltke, Doc. No.20, *Moltkes Militaerische Korrespondenz: Aus den Dienstschriften des Krieges 1870/71: Erste Abtheilung: Der Krieg bis zur Schlacht von Sedan*, (hereafter *MMK*) Moltkes Militaerische Werke, I, Militaerische Korrespondenz, Dritter Theil, Erste Abtheilung (Berlin, 1896), 132. This quote is taken from a document written for the guidance of department chiefs of the Great General Staff and concerns the "forward march" of the army into France, dated 6 May 1870. See comment from the document's editors, same page.

6. Moltke, "Erste Aufstellung der Armee," *MMK*, 116.

7. Ibid., 114.

8. Ibid., 114. The German phrase is 'moeglichst stark und offensiv.'

9. Ibid., 115.

10. Ibid., 115.

11. Doc. No. 20, Ibid., 132. Moltke describes France's capital as being "of greater importance . . . than in other countries." See Graf Helmuth Karl Bernhard von Moltke, *Gesammelte Schriften und Denkwuerdigkeitendes General Feldmarschals Grafen Helmuth von Moltke*, Dritter Band: Geschichte des deutsch-franzoesischen Krieges von 1870-71 (Berlin, 1891), 8.

12. Howard, *Franco-Prussian War*, 246-47, 285, 317.

13. I have chosen to use the term "assembly area" rather than "deployment area." The word "assembly" best describes what was going on in this area; the term "deployment" is better reserved for the process of moving troops from marching to fighting formations. The "Erste Aufstellung der Armee" ("The First Assembly of the Army)—the final version of which was prepared in the winter of 1868-69 and updated through July 1870—was one of Moltke's primary planning documents. See Moltke "Erste Aufstellung," *MMK*, 114-30.

14. Prussian historical sources almost invariably use the term *Aufmarsch* to describe the movement of troops to the assembly area, and *Vormarsch* to describe the movement of the assembled units forward to contact with the enemy. Cf. the language used in Moltke, "Erste Aufstellung der Armee," 114-30, and that used in Moltke, Doc. No. 20, *MMK*, 131-34. An editorial comment (ibid., 31) describes Doc. No. 20 as describing the

Vormarsch of the army.

15. Moltke, Doc. No. 20, *MMK*, 131-34.

16. Ibid., 126, 131-33. The "Erste Aufstellung der Armee," first written in the winter of 1868-1869, underwent a revision. In the first version the Reserve Army was designated the Fourth Army. Since the designation "Reserve Army" was used for these forces in the actual war, all references to the Fourth Army in this Chapter have been transposed to "Reserve Army."

17. Ibid, 132. The strengths of the First and Third Armies are taken from Moltke, "Erste Austellung der Armee," *MMK*, 126. The figures used are planning figures, not the actual wartime figures. The strength of the Third Army is given as either 99,000 or 130,000, depending on whether two Bavarian corps participated. Since they did participate in the war, the larger figure is used in the text.

18. Moltke, *GesammelteSchriften*, 11. The specific reference here concerns a possible French attack on the forces coming out of the defile of the Haardt Mountains. The idea of using the First Army as an offensive flank agrees very well with Moltke's views on strategy, and it may reasonably be considered part of his intentions for this army during the planning stage. See discussion below on the operational goal of attacking an enemy simultaneously in the front and flank.

19. Moltke, Doc. No.20, *MMK*, 132.

20. Ibid., 135.

21. Ibid., 133.

22. Moltke, *Gesammelte Schriften*, 8.

23. Howard, *Franco-Prussian War*, 13-182 *passim.*

24. Moltke, "Erste Aufstellung," *MMK*, 120.

25. Ibid., 120-21.

26. Ibid., 132. The distance Moltke gives for the forward march is 15 miles; but it must be borne in mind that the German mile corresponds to a league, or 3.456 English miles. (This figure is used for all subsequent conversion from the archaic German mile to modern English miles.)

27. *Hammond Family Reference World Atlas* (New York, 1972), 24.

28. Moltke, Doc. No. 20, *MMK,* 132-34.

29. *Hammond Family Reference World Atlas*, 24.

30. Moltke, "Erste Aufstellung," *MMK*, 123.

31. Moltke, "Erste Versammlung der Armee bei einem Kriege gegen Frankreich allein," *MMK,* 105. This was an earlier version of the plan for concentration against France, prepared in 1868 and reworked in January and March 1868. See editorial note, ibid., 98.

32. Moltke, "Erste Aufstellung," *MMK*, 124.

33. Ibid., 123.

34. See discussion below.

35. Howard, *Franco-Prussian War*, 85-223 *passim.*

36. The term "descending spiral" is often used in connection with Colonel John Boyd's observation-orientation-decision-action (OODA) loop concept. See Martin Van Creveld et al, *Air Power and Maneuver Warfare* (Maxwell Air Force Base, AL, 1994), 3.

37. See Grosser Generalstab, Kriegsgeschichtliche Abteilung I, *Moltke in der Vorbereitung und Durchfuehrung der Operationen*, Kriegsgeschichtliche Einzelschriften, Heft 36 (Berlin, 1905), 131. Both Moltke, in the popular history he wrote after the war,

and the German General Staff's official history of the war stated or implied that driving the French away from Paris was an operational goal before the war. See Moltke, *Gesammelte Schriften*, 8; and Grosser Generalstab, Kriegsgeschichtliche Abteilung I, *Der deutsch-franzoesischeKrieg, 1870-1*, Erster Theil: Geschichte des Krieges bis zum Sturz des Kaiserreiches, Heft 1: Die Ereignisse im Monat Juli (Berlin, 1872), 73. But the final draft of "Erste Aufstellung der Armee," and a marginal note in Moltke's own hand, does not mention driving the French forces north away from Paris. Moltke, "Erste Aufstellung der Armee," *MMK*, 114-31. A note from Moltke to his department chiefs concerning the Prussian forward march from the assembly areas specifically says that once Metz had been passed on the left, and an attempt made to cross the Moselle between Lunéville and Pont à Mousson, a decisive battle would ensue "concerning which nothing can be foreseen in detail." Moltke, Doc. No. 20, *MMK*, 131-32. Another General Staff work takes some pains to establish the origins of the operational goal of driving the French north away from Paris. It establishes that Moltke on two occasions in 1867 mentioned bypassing the French to the south on the way to Paris. Apparently he anticipated that the French might defend Paris by means of a flanking position in the neighborhood of Lunéville. In that case he would have a choice of either following the French to the south or bypassing them on the way to Paris. See Grosser Generalstab, *Moltke in der Vorbereitung und Durchfuehrung der Operationen*, 92-93, 99.

38. Moltke, *Gesammelte Schriften*, 8. See also Howard, *Franco-Prussian War*, 246-47, 285.

39. Howard, *Franco-Prussian War*, 183-223.

40. The "Instructions" were substantially written by Moltke, although he did have help. See the editorial note in Graf Helmuth Karl Bernhard von Moltke, "Aus den Verordnungen fuer die hoeheren Truppenfuehrer (1869)," *Moltkes Taktisch-strategische Aufsaetze aus den Jahren 1857 bis 1871*, Moltkes Militaerische Werke, II, Die Thaetigkeit als Chef des Generalstabes der Armee im Frieden, Zweiter Theil (Berlin, 1900), 167-69. The passages quoted in this paper almost certainly stem directly from Moltke.

41. Moltke, "Bemerkungen vom April 1861 ueber den Einfluss der verbesserten Feurwaffen auf die Taktik," *Taktisch- strategische Aufsaetze*, 31-32.

42. Moltke, "Aus den Verordnungen fuer die hoeheren Truppenfuehrer (1869)," ibid., 208.

43. Ibid.

44. Ibid., 196, 210.

45. Ibid., 210-11.

46. Gordon A. Craig, *The Battle of Koeniggraetz: Prussia's Victory Over Austria, 1866* (New York, 1964), 82-83.

47. Moltke, "Aufsatz vom Jahre 1871 'Ueber Strategie,'" Moltkes *Taktisch-strategische Aufsaetze*, 290.

48. Moltke, "Erste Aufstellung," *MMK*, 124-25.

49. Ibid., 121-23, 125.

50. Moltke, "Aufsatz vom Jahre 1859 ueber Flankenstellungen," *Taktisch-strategische Aufsaetze*, 261-62.

51. Moltke, "Erste Aufstellung der Armee," *MMK*, 116-20.

52. Moltke, "Aufsatz vom Jahre 1871 'Ueber Strategie,'" *Taktisch-strategische Aufsaetze*, 291-2.

53. Ibid., 291.

54. Moltke, "Normale Marschtiefe nach dem Kriegs-Verpflegungsetat vom 7. November 1867," ibid., 253. The distance given is 3.75 *Meilen*, which has been converted to English miles using the conversion factor of 3.456.

55. Moltke, "Aufsatz vom 16. September 1865, 'Ueber Marschtiefen,'" ibid., 237.

56. Moltke, Doc. No. 20, *MMK*, 132.

57. Moltke "Aufsatz vom 16. September 1865 'Ueber Marschtiefen,'" *Taktisch-strategische Aufsaetze*, 236.

58. Moltke, Doc. No. 20, *MMK*, 132-34. For Moltke's views on leaving wagon trains behind, see Moltke, "Aufsatz vom 16. September 1865 'Ueber Marschtiefen,'" *Taktisch-strategische Aufsaetze*, 235.

59. Moltke, Doc. No. 20, *MMK*, 135.

60. Moltke, "Erste Aufstellung der Armee," ibid., 120.

61. These figures, with the exception of the figure of 150,000 for the Second and Reserve Armies, are taken from Moltke, "Erste Aufstellung der Armee," ibid. This source shows the Second and Reserve Armies as being 184,000 men strong, which contradicts the figure of 150,000 men given for the Second and Reserve Armies in Moltke, Doc. No. 20, ibid., 133. It seems likely that Moltke used a rule of thumb of 30,000 men for each of six army corps, yielding 150,000 men, rather than going back to more exact computations. The difference is immaterial for the purpose at hand.

62. For instance, see Howard, *Franco-Prussian War*, 83.

63. Moltke, *Moltkes kriegsgeschichtlicheArbeiten: Kritische Aufsaetze zur Geschichte der Feldzuege von 1809, 1859, 1864, 1866 und 1870/71*, Moltkes Militaerische Werke, III, kriegsgeschichtliche Arbeiten, Zweiter Theil (Berlin, 1899), 149-59.

64. Verdy du Vernois, *Studien ueber den Krieg*, Dritter Teil: Strategie, Fuenftes Heft: Einzelgebiete der Strategie, 3. Abteilung: Operationslinien, 2. Unterabteilung: Seit Einreihung der Eisenbahnen in die Kriegsfuehrung (Berlin, 1906), 82-89.

65. Moltke, "Aufsatz vom 16. September 1865 'Ueber Marschtiefen,'" *Taktisch-strategische Aufsaetze*, 236-37.

66. Ibid., 237. The German phrase translated as "the essence of strategy" was "das Wesen der Strategie."

67. Howard, *Franco-Prussian War*, 83-85, 89.

68. Howard says that this phenomenon was characteristic of all the great battles of the campaign. See ibid., 103.

69. Ibid., 130, 145.

70. Eberhard Kessel, *Moltke* (Stuttgart, 1957), 593-763 *passim*.

71. Moltke, *Gesammelte Schriften*, 6.

72. Carl von Clausewitz, *On War*, edited and translated by Michael Howard and Peter Paret (Princeton, NJ, 1976), 177.

73. Moltke, "Aus den Verordnungen fuer die hoeheren Truppenfuehrer vom 24. Juni 1869," *Taktisch-strategische Aufsaetze*, 206.

74. Moltke, "Aufsatz vom 16. September 1865 'Ueber Marschtiefen,'" *Taktisch-strategische Aufsaetze*, 237.

4

Two Views of Warsaw: The Russian Civil War and Soviet Operational Art, 1920–1932[1]

Jacob Kipp

Over the last decade Western military historians and soldiers have come to appreciate the enduring contributions of Soviet military science to the conduct and study of war at the operational level, that is, at echelons above corps and on the scale of theater-strategic campaigns.[2] As James Schneider has pointed out, the U.S. Army's approach to the theoretical implications of operational art in the 1980s owed much to the debate and discussions among Red Army commanders in the 1920s.[3]

Today, many scholars acknowledge the vitality and creativity of Soviet military theory in developing such concepts as deep battle, deep operations, and successive operations. A number of recent accounts demonstrate the number of young Red commanders and Tsarist military specialists who contributed to the development of operational art and laid the foundations for Soviet military strategy in the interwar period.[4] From the late 1930s until the 1970s most Soviet military historians tended to emphasize the infusion of Marxism-Leninism into Soviet military science as the dominant and decisive theme in the development of Soviet military theory in the interwar period and to downplay the significance of Stalinization. Later events, particularly the politicization of military theory and attacks upon Tsarist-trained military specialists (the *voyenspetsy)*, the blood purge of the military, the Cult of Stalin,[5] and the manufacture of an entire pseudo-history of the Civil War, conspired to rob the Red Army of its past, obscure the origins of operational art, and plant seeds of confusion and uncertainty about the contribution of individuals to the development of operational art in the interwar period. Many of the most important contributors to these developments were labeled "enemies of the people," imprisoned, liquidated, and then transformed into "nonpersons." With their liquidation the Soviet Army lost much of its own past. This situation has greatly handicapped the study of the origins and development of operational art.

In the West what brought back a focus on campaign planning was the Soviet military challenge in Central Europe. But in the process of studying the Soviets as probable adversaries, some Western military analysts embarked on a process akin to dialectical synthesis, that is, studying Soviet concepts, critiquing them and then creating a synthesis that pushed theory forward. The late Brigadier Richard Simpkin analyzed the continuing relevance of deep operations to the conduct of theater war and outlined Marshal Mikhail Nikolaevich Tukhachevsky's contribution to its development.[6] Simpkin argued that Tukhachevsky's approach to the development of operational art contributed to a distinctly Soviet approach to the military-technical side of doctrine that stressed combined arms and the need for theory to lead technology. This approach he argued still had relevance for military analysts struggling to come to grips with combat in the postnuclear era.[7]

While accepting a leading role for Tukhachevsky in the development of Soviet operational art in the interwar period, the current chapter will place that development within the larger context for the formulation of Soviet military art and military science during the first decade of the Red Army, 1918-1928. To understand the Red Army's theoretical contribution to the development of operational art the reader must see those events in context: the creation, formation, characteristics, and initial combat experience of the Red Army, and, especially, the role of military specialists (*voyenspetsy*) in making the legacies of the Imperial Army, General Staff and Academy of the General Staff available to the Red Army, and the Academy of the General Staff.

The evolution of the Red Army as a combat force and the issues raised by one of the final campaigns of the Civil War, "the Campaign beyond the Vistula," that is, Tukhachevsky's advance on Warsaw, played a very special role; it contributed to the Red Army's efforts to address the questions of future war and operational art by assessing combat experience in the First World War and the Civil War. Lenin viewed the capture of Warsaw in 1920 as the next step in the world revolutionary process. The destruction of the new Polish state would rock the Versailles system and its *cordon sanitaire*. The Red Army had the chance to carry the revolution up to the very borders of a defeated and unstable Germany, where a powerful working class could give the world revolutionary process new mass and momentum.[8] The severe rebuff of Tukhachevsky's Red offensive dealt by Polish marshal Pilsudski thereafter fueled a lively debate within the Red Army over methods and leadership.[9] Defeat before Warsaw in August 1920 forced the Red Army to serious "rethinking, retraining, and rearmament" during the interwar period.[10] The assessments of the Reds' campaign failure greatly informed the post-Civil War search for a new, unified military doctrine. This search resulted in an uneasy but fruitful amalgamation of Tsarist theory and Soviet practice, and laid the intellectual foundation for Soviet operational art.

In considering Warsaw this chapter will address what were the two dominant paradigms for understanding that campaign within the Red Army: (1) that there was a unique military art and doctrine for the "Red Army as a revolutionary force" and (2) that Warsaw was a problem of the evolution of military art, connected to the experience on the Eastern Front during World War I, informed by relevant military experience from the Civil War, and pointing toward new topics and problems associated with future war.

Tukhachevsky's essay *The Campaign Beyond the Vistula*, first given as a lecture to the advanced course at the Academy of the General Staff in Moscow in February 1923, was part of that process.[11] He wrote as both the field commander of the forces who conducted that offensive and as the commander of the Western Military District charged with working out the plans for the conduct of initial operations in that theater in a future war with Poland. Tukhachevsky was a theorist who used combat experience to study the evolution of military art. Like the British writer Major General J.F.C. Fuller, Tukhachevsky embraced military science, and similarly emerged as one of the champions of mechanized warfare and the armored forces. Like Fuller, whom he read, Tukhachevsky constantly engaged in polemical struggles on the changing nature of military art.[12] He emerged as the most determined champion of a strategy of "annihilation," a philosophy for which Fuller roundly denounced him. As Tukhachevsky wrote elsewhere in 1923, the primary military objective of armed conflict is the complete destruction of the opposing army. "An operation is the organized struggle of each of the armies for the destruction of the men and materiel of the other. Not the disruption of some hypothetical, abstract nervous system of the army, but the destruction of the real organism—the troops and real nervous system of the opponent, the army's communications, must be the operational goal."[13] His approach to the conduct of operations, as befitting his youth, temperament, and experience, was tactical in focus. Tukhachevsky pitted his own concept of operational art into this strategy of annihilation and entered into a debate with those who sought to adapt operational art into a strategy of attrition. In the ensuing debate the competing interpretations of the Warsaw campaign had a profound impact on the development of a Soviet concept of operational art, which emerged in the late 1920s.[14]

TUKHACHEVSKY AND THE RUSSIAN CIVIL WAR

With Marxist revolutionaries in power in Moscow after the Bolshevik Revolution of 1917, Warsaw became the point where revolutionary politics and military art came together. As the chief military threat to the Soviet state, Poland assumed a special place in the political-military and military-technical debates of the 1920s. Tukhachevsky stood at the center of those debates, and his changing role is closely connected to the origins and evolution of the operational art. Tukhachevsky, although born of a gentry family and a Guards officer, became the symbol of the young Red commander. Serving briefly at the front in the First World War, he was captured by the Germans and held with other Allied officers as a POW. For his repeated attempts to escape he ended up in Ingolstadt fortress in upper Bavaria. But even in this strict-regime prison his attempts to escape continued. Successful on his fifth attempt, he crossed the border into Switzerland. Making his way back to Russia, Tukhachevsky soon found himself in Petrograd. While in captivity Tukhachevsky had expressed his support for the February Revolution that overthrew the Tsarist regime. On his return to Russia his sympathy went with the radicals, especially Lenin's Bolsheviks. In the spring of 1918 Tukhachevsky began to work for the All-Russian Central Executive Committee of the Soviets.[15] Raising rapidly from junior officer to army and front commander, in 1919 Tukhachevsky led the Red Army's Eastern Front in its counteroffensive operations against Admiral Kolchak's White forces, which controlled the Urals and much of Siberia. Tukhachevsky also led the Red Army's campaign into Poland in the 1920s, and his own appreciation of the military art was shaped by his experience during the Civil War. This not only shaped his campaign strategy in Poland but also contributed to his interpretation of those events, which became a critical part of ongoing and lively debate about its significance. Tukhachevsky saw revolutionary élan as a new force to reshape the military art.

The Soviet experience in the Civil War proved qualitatively different from that of the First World War on either the Western or the Eastern Fronts. If the Imperial Army had suffered from the economic backwardness of old Russia, the Red Army had to confront the utter disintegration of the national economy. Revolution, civil war, international boycott, and foreign intervention combined to undermine national economic life. The regime's response, War Communism, was less social utopia and more a form of barrack socialism, in which all resources were organized to field a mass army equipped with the most basic instruments of industrial war—the rifle, machine gun, and field artillery. And even in the procurement of these vital weapons the level of production fell radically in comparison with what had been achieved by Russian industry during the First World War. Thus, in 1920 the production of rifles was three times less than in 1917.[16] Unlike the Whites, the Red Army remained reliant on captured stocks of Tsarist arms and supplies. These materiel limitations notwithstanding, by the end of the Civil War the Red Army fielded a ragtag force of 5.5 million.[17]

The Civil War was also noteworthy for a number of politico-geostrategic features, which had a profound impact on the nature of the struggle. First, it was in every sense a civil war in which neither side asked nor gave quarter. The Russia over which the Reds, Whites, and Greens struggled might be described as a few island cities in a sea of peasant villages. The cities emptied as the links between town and countryside collapsed. Red Guard detachments swept through villages seizing grain and recruiting soldiers. Red terror and White terror mounted in scale and intensity. At times it was difficult to distinguish between combatants and brigands. The Red and White armies were notoriously unstable with a persistent problem of desertion. In 1920, when he was preparing for the Western Front's offensive, Tukhachevsky had to face the fact that the Commissariat of War could not find many additional troops to support the operation, and so he instituted a campaign to extract 40,000 deserters from the region's villages and back into service. Within a month Western Front found that it had "extracted" 100,000 deserters, whose presence taxed the supply and training capacity of the front.[18] Such reinforcements were unreliable in the attack and tended to vanish at the first sign of disaster.

The second reality of the Civil War was the fact that the Bolsheviks controlled the central heartland around Moscow and managed to maintain an effective, if much reduced in scale, rail system, which permitted them to use their internal lines of communication to great effect. On the other hand, the White armies fought on the periphery of Russia, in lands often inhabited by non-Russians who had no great interest in the revival of a centralized Russian state. The presence of the White armies on the periphery, especially in south Russia, the Kuban, and Siberia, meant that operations were frequently conducted in "underdeveloped [*malokul'turnye*] theaters of military action." As R. Tsifer observed in 1928, the Civil War seemed to confirm the general rule that the more developed the theater of war, the more likely the emergence of positional forms of warfare, and conversely, the less developed the theater of war, the greater the opportunities for the employment of maneuver forms of combat.[19] This situation, when linked to the low density of forces, the ineffectiveness of logistical services, and the combat stability, created conditions for a war of maneuver. It was not uncommon, as Tukhachevsky pointed out, to have each side launch operations that would sweep 1,000 versts (600 miles) forward and another 1,000 versts back.[20] The instability of the rear in military and political terms meant that a successful offensive, if a vigorous pursuit could be maintained, would often lead to the routing of the opponent and the disintegration of his political base.

Maneuver in this case took the form of a "ram" of forces aimed directly at the enemy in the hope of disorganizing and demoralizing him. It would be fair to characterize this operational approach as an attempt to substitute mobility for maneuver, since the Red Army lacked both the staff assets and communication facilities to sustain the necessary command and control to carry out more complex maneuvers which might lead to the encirclement and destruction of enemy forces.[21] In Tukhachevsky's case this approach was linked with the concept of political subversion and class war as a combat multiplier, what he called "the revolution from without,"[22] wherein an "offensive of the revolutionary army of the working class into a neighboring bourgeois state can break the bourgeoise's power there and transfer power to the dictatorship of the proletariat."[23]

This approach demanded increasing the mobility of the Red Army. Russian cavalry had not distinguished itself particularly during the First World War. Now under Civil War conditions, cavalry recovered its place as the combat arm in a war of maneuver. The loyalty of the Don Cossacks and the support of many senior cavalry commanders gave the Whites substantial initial advantages in the use of cavalry.

Trotsky's famous call "Proletarians to horse!" initiated the process of creating a "red cavalry."[24] Soviet cavalry units were raised from the beginning of the war; however, greater attention was paid to creating troop cavalry detachments to provide the eyes and security screens for the newly formed infantry divisions. Army cavalry, that is, cavalry units organized into independent brigades and divisions, were gradually formed into corps and later armies.[25]

The raid mounted by General K. K. Mamontov's cavalry in August-September 1919 provided the stimulus for the creation of the legendary Red Cavalry army, the Konarmiya. In order to take pressure off General Anton Denikin's White forces, Mamontov's Forth Don Cavalry Corps (7,500 sabers) undertook an independent raid deep into the rear of the Southern Front. The Thirty-sixth and Fortieth Divisions, which held the 100 kilometer section of the line through which Mamontov's corps passed, were widely dispersed, and Mamontov used air reconnaissance to find a sector where his cavalry could slip through without serious opposition. Using his air reconnaissance to avoid contact with Bolshevik units, Mamontov struck deep into six gubernias,[26] wrecking the railways and destroying military stores as he advanced.[27] The Military Revolutionary Council, the Revvoyensovet of the Republic, took this threat seriously and created an internal front under the command of M. M. Lashevich to deal with Mamontov's corps. On its return to Denikin's lines the corps's pace slowed under the weight of booty, and Lashevich was able to concentrate Red forces against its strung-out columns. Mamontov reached Denikin's lines but suffered serious losses on the retreat south from Kozlov to Voronezh.[28] The use of air assets to provide effective reconnaissance for large-scale cavalry raids was noted by the Red Army and became an important part of its own concept of strategic cavalry.[29] Intelligence Units and Counter Intelligence Organs quickly grasped the military and political effects of such raiding maneuvers. The Eighth Red Army had been totally routed, a general panic had been created in the Soviets' rear area, and the most strenuous military and political measures were required to deal with the threat posed by Mamontov's raid. These included the systematic use of Red terror and the Cheka internal security police.[30]

In November the Revvoyensovet ordered the creation of the Konarmiya under the command of S. M. Budennyi, a former NCO in the Tsarist army and then the commander of the First Cavalry Corps. Konarmiya was initially composed of three cavalry divisions, an armored car battalion, an air group, and its own armored train. Later, two other cavalry divisions were added and an independent cavalry brigade was also included.[31] The basic units of the Konarmiya were its cavalry divisions, armed with rifles, sabers, revolvers, and hand grenades. Each division was also to have 24 machine guns mounted on *tachanki* (carriages), but in practice the number was often two or three times higher.

Budennyi's Red Cavalry quickly became the stuff of legends. Issac Babel, who served as a political commissar with one of its units, immortalized its exploits in a cycle of short stories.[32] The legend later turned into official myth as Budennyi, Stalin, and others invented history to fit their cults of personality. In the decade after the Civil War it was still possible to give a reasonably objective evaluation to the contribution of the Konarmiya and strategic cavalry in general to Soviet operations on the various fronts of the Civil War. Strategic cavalry repeatedly played the role of a shock force, striking deep into the enemy rear, disrupting command and control, and demoralizing forces. Among the most celebrated of these operations were those in the Ukraine in June-July 1920, when Konarmiya was redeployed from the Caucasian Front to the Southwestern Front to form the strike group for a drive to liberate Kiev and push the Poles out of the Ukraine. At the start of the operation, Budennyi's Konarmiya had 18,000 sabers, 52 guns, 350 machine guns, five armored trains, an armored car detachment, and eight aircraft. The Polish Third Army was spread thin and had few effective reserves. Thus, one cavalry division was able to break through the lines and mount a raid on Zhitomir-Berdichev in the first week of June. The Polish commander responded by shortening his lines and giving up Kiev. The blows of the Konarmiya were in this case combined with pressure from the Soviet Twelfth Army, and this created the impression that the Polish defenders faced the possibility of being surrounded and cut off.[33]

Budennyi's force engaged in 43 days of intensive combat without effective logistical support. Cavalry brigades which at the start of the campaign had numbered 1,500 sabers were down to 500 or less by the end of the fighting. The fighting at Zhitomir and Rovno exemplified the combined arms approach which typified Soviet employment of strategic cavalry. It also showed its limited ability to engage in sustained combat.[34] At the same time, the Zhitomir and Rovno operations exemplified the psychological impact of the strategic raiding force. Marshal Pilsudski credits Budennyi's Konarmiya with an ability to create a powerful, irresistible fear in the deep rear. Its effect on the Polish war effort was like the opening of another, even more dangerous front within the country itself.[35]

TUKHACHEVSKY AND THE CAMPAIGN BEYOND THE VISTULA

The Red Cavalry's success at Rovno set the stage for one of the most controversial and frequently studied operations of the Civil War, that is, Marshal Tukhachevsky's general offensive of July-August 1920, in which his Western Front struck beyond the Vistula to threaten Warsaw. Pilsudski's counterattack, coming at the very gates of Praga and resulting in the destruction of major Soviet formations pinned against the Polish-East Prussian border, became known as the "Miracle of Warsaw." On the banks of the Vistula the romantic dream of carrying the world revolution on the bayonets of the Red Army died. Tukhachevsky's "ram" pushed the Polish army back but did not break it. No massive revolutionary uprising erupted in the Polish rear. Indeed, Poles, who had lost their independence in the eighteenth century and been subjected to Russian military interventions in the nineteenth century, rallied to the cause of national independence, rather than display the class solidarity of the world revolution.

More realistic Soviet military assessments of the campaign said that the "miracle" was that the bedraggled, unfed, poorly armed, ragtag divisions of the Western Front got as far as they had. Tukhachevsky's general offensive took place without adequate reserves, effective command and control, or logistical support.[36] General Maxime Weygand, who was in Warsaw as an adviser to the Poles during their counteroffensive, said of Tukhachevsky's forces: "Elle souffre des faiblesses de toute armeé improvisée."[37]

Believing his own theory about "revolution from without," Tukhachevsky fell into the trap of assuming that the psychological weight of the advance would break the will of the Polish defense without having to destroy those forces in the field. His forces did manage to push the Polish defenders back over several natural defensive positions and the line of German emplacements along the Auta.[38] But the advance exhausted the troops and carried these troops beyond their own supplies. With a direct order from Commissar of War Trotsky to capture Warsaw as quickly as possible, Tukhachevsky's Western Front embarked upon what its commander labeled "the decisive offensive," the envelopment of the Polish capital from the northwest of Modlin, seeking to cross the Vistula on a broad front with about one-half the manpower of the defending Poles.[39] Tukhachevsky counted on pressure from the south on the Lublin axis to divert Polish forces in that direction. But First Cavalry Army was tied up in fighting around Lvov, not Lublin. Pilsudski's counterattack struck the over-extended forces of Western Front near Siedlice and drove a wedge between Tukhachevsky's Thirteenth Army and the Mozyr Group. The attack threw RKKA (Red Army) Western Front back in disarray and trapped the Fourth Army against the East Prussian border.[40]

The geographic peculiarities of the theater, that is, the fact that Belorussia and the Ukraine, taken as a whole, are bisected by the Pripyat Marshes, created two distinct axes of advance toward the Vistula. The existing Soviet command structure called for Tukhachevsky's Western (Belorussian) Front to direct the fighting north of the Pripyat Marshes and A. I. Egorov's Southwestern Front (Ukrainian) to direct the fighting south of the Pripyat Marshes. This structure had made strategic sense until the Poles fell back on the Vistula and Bug Rivers and the front line advanced west of the Pripyat Marshes. Now this military case of "dual power" combined to frustrate Soviet control of the Vistula campaign. In addition to directing the fighting in the Kiev sector, Southwestern Front also had to combat Wrangel's army based in the south and cover the potential threat of Rumanian intervention. Budennyi's Konarmiya persisted in its attacks toward Lvov, even after S. S. Kamenev as commander in chief had ordered it and the Twelfth Army to regroup, join Western Front, and undertake a drive toward Lublin to relieve pressure on Western Front. Southwestern Front Commander Egorov found himself caught trying to manage operations on two axes without staff support and did not feel "the beating pulse of the operation."[41] Thus, Tukhachevsky's Western Front lacked support from the south when its Fourth, Fifteenth, and Third Armies tried to turn Warsaw from the north by crossing the Vistula between Modlin and Plock. This allowed Pilsudski to carry out a regrouping of forces south of Warsaw and to prepare his counteroffensive against Tukhachevsky's weak left.

Since Joseph Stalin served as the political commissar of the First Cavalry Army, Budennyi's independence and insubordination became entangled in the political struggles following Lenin's death. Under Stalin's cult of personality the unpleasant truth about Lvov and Warsaw was covered up by blaming Trotsky, the Commissar of War, for ordering the regrouping of forces to support a drive on Lublin. By then Tukhachevsky had been executed by Stalin as a traitor and enemy of the people.[42] Tukhachevsky's public explanation for the defeat in 1923 probably contributed to his fate. Tukhachevsky put the blame on poor intelligence, bad communications, poorly trained commanders and staffs, and strategic misdirection, that is, the failure of First Cavalry Army and Fourteenth Army to change their axis of advance from Lvov to Lublin.[43] Debate over these matters proved intense.

THE *VOYENSPETSY* AND MODERN WAR

Lenin and Trotsky had championed the recruitment of military specialists into the Red Army in the face of serious opposition from the left radicals in their own party, who favored a militia-style army and partisan warfare.[44] The Red Army relied heavily upon Tsarist military specialists for combat leadership, staffing, and training. By the end of the Civil War about one-third of all Red Army officers were *voyenspetsy* and in the higher ranks the ratio was even greater. Thus, 82 percent of all infantry regiment commanders, 83 percent of all division and corps commanders, and 54 percent of all commanders of military districts were former Tsarist officers.[45] *Voyenspetsy* from among the faculty and staff of the Nikolaev Academy of the General Staff dominated the Red Army Main Staff and the supporting staff of the Revvoyensovet of the Republic.

In the early and mid-1920s, the debate over the Polish campaign's significance for the operational art influenced the minds of those young Red Army officers, who formed a Soviet military intelligentsia. Thanks to their combat experience during the First World War and the Civil War these officers were able to combine practical experience and a general staff education. They sought to create a unique and ideologically correct Soviet military science and doctrine.[46] Before Stalin's terror stilled debate many Soviet officers believed it possible to promote the development of military theory through an open and active debate. Aleksandr Ivanovich Verkhovsky (1886-1938), officer of the general staff, military historian, veteran of the Russo-Japanese War and First World War, Minister of War in the Provisional Government in September-October 1917, and *voyenspets* from 1919, described this collaboration in the 1920s when he taught tactics at the Military Academy of the Red Army as a "war on two fronts." This prescription marked a struggle among proponents of three competing visions of the army: the conservatives, who wanted to maintain past views because they were sanctioned by history and the unchanging laws of military science; the realists, who saw the need for change as dictated by objective conditions and the demands of future battlefields; and the futurists, who on the basis of their experience in the Revolution and Civil War put their faith in crude military means and political agitation and trusted in class struggle to ignite revolutions behind the enemy's lines. In looking back on this struggle in the area of tactics he concluded that it had been one full of vitality and even joy. The Red Army had made startling progress.[47]

Similar progress was made in the realm of "higher tactics" or "lower strategy," as studies of the operational level of war were known at the Military Academy of the RKKA in the period 1918-1923. Pre-revolutionary Russian military writings had already criticized those who sought to direct this entire process toward a single decisive operation. A. A. Neznamov noted that in the last century there was only one example of such an operation—Napoleon's 1806 campaign against Prussia. A number of these works were republished.[48] Short wars and rapid decision might be the objective of both sides and even be in the interests of both, but the complexity of the task made this goal very difficult to achieve. However, the initial operation created a new situation and imposed on the commander and his staff the need to plan and conduct further operations according to the new circumstances, trying to seize the initiative or playing for time. While rejecting the idea of directing all efforts toward a single, decisive engagement, these works affirmed the need to link each battle into a coherent whole in keeping with the campaign plan.[49]

One of the first to take issue with Tukhachevsky's interpretation of the Polish campaign was the *voyenspets* Boris Miklhailovich Shaposhnikov, who had been chief of the Operations Directorate of the Field Staff. Shaposhnikov responded to Tukhachevsky's essay with his own analysis of 1920, offering a point-by-point analysis of the campaign from the May offensive to the Polish counter-offensive of August. Shaposhnikov, who went on to write what would become the classic Soviet work on the role of the general staff in peace and war, *The Brain of the Army*, took issue with Tukhachevsky on many points, but he returned to the question of campaign planning. There he pointed out two serious flaws: the overevaluation of the impact of revolutionary unrest on the opponent and the failure to take into account the fact that the campaign unfolded without the annihilation of Polish forces in the initial operations. Tukhachevsky's "ram strategy" had thrown the Poles back from the Bug to the Vistula, but failing to achieve culmination, an exhausted Red Army had to face a Polish force strengthened by new reserves. Drawing on Clausewitz's *On War* to support his points, Shaposhnikov saw the chief failing in Western Front's planning of its offensive operations to have been a miscalculation of the forces available and their offensive capabilities. The issue was not politics or strategy, but a problem of a faulty campaign which did not take into account the culminating point of the campaign and had gone beyond it. On the key issue of command and control Shaposhnikov accused Tukhachevsky of trying to resolve the issue by reverting to Napoleonic "handicraft" in the face of mass war. While Tukhachevsky spoke of inadequate technical means to expedite command and control and "the unpreparedness of subordinate chiefs," Shaposhnikov pointed to a more serious structural problem in developing a command and staff system that would assure effective strategic and operational direction from strategic headquarters, through army groups, to armies, and so on.[50] Shaposhnikov warned against a pedantic and amateurish approach to the study of the 1920 campaign which sought to set blame and called for a more objective one. Shaposhnikov had discussed the campaign beyond the Vistula in Clausewitzian terms, but his book might best be described as an anti-memoir to Tukhachevsky's own memoir. A more sustained critique of Tukhachevsky came from another *voyenspets*, who radically altered the terms of the discussion by introducing a new category into military art.

A. A. SVECHIN ON OPERATIONAL ART AND STRATEGY

This opposing theoretical position belonged to General-Major Alexander Andreevich Svechin (1878-1938), who first applied the term "operational art" (*operativnoe iskusstvo*) to refer to a third category of military art between strategy and tactics.

In the same year that Tukhachevsky published his account of the campaign beyond the Vistula, Svechin coined the term "operational art," in a series of lectures on strategy in 1923-1924 at the Military Academy of the RKKA.[51] He described operational art as the bridge between tactics and strategy, i.e., the means by which the senior commander transformed a series of tactical successes into operational "bounds" linked together by the commander's intent and plan and contributing to strategic success in a given theater of military actions.[52] Over the next several years Svechin turned these lectures into a book, *Strategy*, which first appeared in 1926. Whereas Tukhachevsky had come at Warsaw from the perspective of the field commander, moving from the tactical to the strategic, Svechin began his analysis at the strategic level of war and moved to the operational. Tukhachevsky wrote as a committed revolutionary, while Svechin brought an analytical approach to military science.

Svechin, a prominent Tsarist military intellectual, was one of the first senior Tsarist officers to join the Red Army as a military specialist. Svechin, as a *voyenspets*, brought with him considerable combat experience and well-developed ideas on the nature of modern war and the evolution of military art. During the First World War, Svechin served at Stavka, then commanded a regiment and division, and from September 1917 was chief of staff of the Northern Front. Following the October Revolution and the disbandment of the Imperial Army, Svechin joined the Red Army of Workers and Peasants in March 1918 and held a series of posts connected with the defensive "screens" which the Soviet regime attempted to maintain along the front while it negotiated peace with the Central Powers. In August 1918, as the Civil War was intensifying, Svechin was appointed chief of the All Russian Main Staff and held that post until October of that year. Thereafter, he took up his teaching duties in the newly established Academy of the General Staff of the RKKA. Svechin served, fought, studied, and wrote in a time of momentous changes in the nature of war. His career as an officer of the Imperial General Staff (*genshtabist*) and Soviet military specialist underscores the themes of continuity and change in the Russian/Soviet military. If there is one theme that unites all of his studies it is those trends which were guiding the evolution of military art under the impact of the industrialization of warfare:

> The great commanders, as with all successful practitioners, were first of all sons of their age. In the epoch of Napoleon the techniques of Frederick the Great were utterly defeated and now the application of the techniques for the Napoleonic epoch lead only to failure. Successful action most of all must be proper to its place and time, and therefore it must agree with the contemporary situation.[53]

His own approach to military art and theory could be described in the same fashion in which he characterized the German military historian Hans Delbrück: a combination of the Hegelian dialectic with historical materialism.[54] Introducing such a dialectical approach to an evolving military art had the same impact on military theory that Einstein had on Newtonian physics. In place of certainty and eternal laws in military affairs there appeared the principle of "relativity" (*otnositel'nost'*) negating the very "decisiveness, absence of vacillation, and goal-directedness," which had so much importance.[55] Svechin emphasized the evolution of military art and warned against any effort to create closed systems on the basis of past combat experience. The proper topic of military history was the study of those tendencies shaping future war.[56]

Svechin combined extensive combat experience in Manchuria and the Eastern Front with a solid mastery of military history and theory. Lenin's government found the Tsarist General Staff's post-1905 approach to the study and use of military history worthy of emulation. One of the first acts of the Soviet Republic in 1918 was the creation of the Commission for the Study and Use of the Experience of the War, 1914-1918.[57] This effort drew upon the talents of many former officers of the Russian General Staff, including Svechin, who headed and provided editorial direction to the project. In the first volume of essays published by the Commission Svechin used the introduction to the volume to call for further study of changes in strategy and tactics made evident by the World War. Regarding the deeper political and socio-economic changes wrought by the world war, Svechin consigned their study to the realm of the Socialist Academy and identified the Commission's work as narrowly military and immediately practical. He recognized the twin problems of masses of information and the need for an operational focus.[58]

Svechin's treatment of the war was noteworthy for the absence of a Marxist analytical framework and the presence of an integral Russian nationalism linking together, even in 1919, the past accomplishments of Russian arms and national military valor, which he described as "a cement, uniting us into one whole."[59] At the same time Svechin promised an objectivity which transcended even that of the Elder Moltke's injunction to his General Staff before writing the history of the Franco-Prussian War: "the truth, only the truth, but not all the truth." Instead, Svechin said that the Commission's motto would be Clausewitz's: "the truth, only the truth, the whole truth." The reputations of commanders from an army overthrown by social revolution did not need the same special care as those linked to an ancient dynasty.[60] Later, when the Commission's task was extended to the study of the Civil War, it proved difficult for Soviet military authors to live up to this standard when studying the RKKA's own experience. A little over a decade later Stalinism made a mockery of even Moltke's formula by substituting outright lies for historical judgment to create its own mythical past and by applying terror transforming historical actors into nonpersons and historical events into nonevents. Yet, for a decade Svechin's standard did remain the criterion for RKKA studies over a wide range of topics. Their high caliber and professional quality owed very much to the example he set.

Svechin's approach to military history was anything but dogmatic. He understood that his views had been shaped by the experiences of his own generation of General Staff officers. He was sympathetic to the young Red commanders, who upon arriving from the fronts of the Civil War, questioned the applicability of school solutions and textbook military science to their war. Svechin noted that these students were soldier-revolutionaries and not traditional student officers. Young men, just arriving from the fronts of a bloody and bitter civil war, were already hardened veterans, having seen combat in the First World War and the Civil War. Full of enthusiasm for a cause but distrustful of the professors from the Tsarist Nikolaev Academy of the General Staff, who were suspected as "class enemies," they refused to be intimidated by classical authorities or to accept the "school" solutions. Their test of instruction was its relevance to their own practical experience in the field. Svechin could see in the face of each man "an idea which is blasphemous to the temple of science, that is, to bring in something of his own—to criticize thoroughly the ideas presented to them. Their enthusiasm merged with a scorn for the old forms of military science."[61]

He came to the operational art with the perspective of a strategist who had thought long and hard regarding the nature of modern war and the geostrategic dilemmas confronting a backward Russia in the age of industrial war. His struggle with Napoleon's épigones at the Nikolaev Academy had led him to emphasize the evolution of military art and made him a close ally of those reformers who saw Russia's defeats in the Russo-Japanese War to have been a result of the General Staff's inability to master modern war. Those reformers, especially Lieutenant Colonel A. Neznamov, had emphasized an operational focus in the development of a military doctrine for the Russian army prior to the First World War.[62] For Neznamov, the Russian defeats in the Far East had one basic cause: "We did not understand modern war."[63]

Svechin's interest in the conduct of operations evolved out of a systematic critique of the failure of tactics to address the problem of troop control in modern theater warfare, first observed during the Russo-Japanese War. Two decades before the term "operational art" was coined reform minded Tsarist officers had noted that modern war had destroyed the symmetry of the Napoleonic paradigm in which tactics were the management of forces on the field of battle and strategy the maneuver of forces to the field of battle. For these officers of the Imperial General Staff Manchuria had been the classroom and the Japanese army the harsh teacher. The Battle of Mukden in January 1905 dwarfed Borodino in firepower, area, and time, and posed a host of new problems relating to the control of troops.

At Mukden three Russian armies, numbering 300,000 men, 1,475 field guns, and 56 machine guns, faced five Japanese armies, numbering 270,000 men, 1,063 guns, and about 200 machine guns. The fighting lasted for six days and covered a front of 155 kilometers and a depth of 80 kilometers.[64] The battlefield had become more vast, less dense, but more lethal. Railroads could move greater masses of troops over greater distances and sustain the flow of men and materiel into the theater. Troop control on this expanded battlefield had become far more difficult as multiple armies operated over broader frontages, raising a host of issues associated with the evolving nature of the application of combined arms to achieve success. The magazine rifle, quick-firing field gun, and machine gun had altered the relationship between offense and defense and called into question the means by which commanders sought to conduct maneuver, fire, and shock. Mass armies, industrialization of society, and the acquisition of new weapons had brought these changes in the scale, physical dimensions, and temporal character of modern combat, replacing the great culminating battle with a series of tactical engagements united by a commander's concept to form a single operation. Successive operations in a theater according to a unified theater conception became a campaign strategy.

Successive operations recast the problem of logistical support in the theater of military actions and raised but did not resolve the problem of pursuit and exhaustion. This operational focus became a means of drawing the attention of senior officers to the need to provide effective leadership over a battlefield, which had been recast in terms of time, space, and scale of combat assets engaged. This battlefield required that control be exercised through a modern headquarters and staff, linked with the front and rear by telegraphic and telephonic ties. Effective troop control called for an effort to link together a succession of tactical bounds under a unified campaign plan designed to achieve strategic success in a theater.

Svechin's analysis addressed those problems, which went beyond the "genius" or lack of genius of a particular commander, in this case the defeated Russian commander, General Aleksei Nikolaevich Kuropatkin. Kuropatkin had been an excellent chief of staff to General Skobelev in the Balkans, had written extensively on that experience, and had later campaigned effectively in Central Asia.[65] As War Minister he directed Russia's rearmament in the years before the outbreak of war and proved a talented logistician. Russia mobilized a half million men and sent them over 5,000 miles by rail. Kuropatkin was a devoted disciple of General G. Leer, who during his long tenure in the Chair of Strategy at the Academy of the General Staff had made Napoleon's campaigns the model. Kuropatkin's initial deployments and the slow buildup of his operations on the Mukden-Port Arthur axis were clear proof that he understood and was applying the concept of the operational line. What he and his staff could not do was provide effective command and control of his forces in the field. The Russian high command spent the entire war in Manchuria seeking the single set-piece battle which would decide the campaign. Its elaborate march-maneuvers to position forces favorably for a general engagement were frustrated by Japanese preemptive meeting engagements. Svechin warned against any cavalier assessment of the lessons of the war,

> in which the failures are explained exclusively either by the inabilities of individual commanders or the super-natural combat capabilities of the enemy, or illiteracy of the Russian peoples, or the unrest within the state. We do not need criminals or idols; they only interfere with the assessment of our mistakes and a rational correction of them.[66]

The Japanese, using German mission-oriented tactics, seized the initiative, threatened Kuropatkin's flanks, and repeatedly forced him to abandon the field after a spirited but inconclusive defense. The Japanese commander, rather than waiting to deploy his forces and then enter into a general engagement, allowed his troops to engage the enemy from the march, thereby seizing the initiative and frustrating Kuropatkin's elaborate plans.[67] Japanese junior officers understood their commander's intent, responded to unexpected developments by exercising their own initiative, and accomplished their tactical missions. That spirit was too often lacking in Russian officers, including officers of the General Staff, who fell back upon school solutions and became "operational lawyers" and bureaucrats, not soldiers. Their first concern was to make sure that no one could question their decisions.[68]

In Manchuria the battlefield assumed a breadth and depth which had been unthinkable only a half century before. It required a new sort of commander who could conquer space and time to bring about concentration of combat power at the decisive point and time to press the combat to culmination. Repeatedly, Japanese commanders achieved such results against numerically superior Russian forces. At Mukden Russian reserves found themselves marching from one side of the battlefield to the other and either taking no decisive part in the action or being so exhausted by the process that they lost their effectiveness. Having lost the initiative to the Japanese, Kuropatkin repeatedly found himself on the tactical defense and forced to withdraw under strong enemy pressure.

Critics such as Svechin concluded that the impact of technology on the scale of battle was in the process of working a radical change in the conduct of war. Russian officers began to speak of a new focal point in military art between strategy and tactics, war and battle. Svechin's forebearer from the Tsarist staff had developed a new terminology to give expression to this intermediary level of combat. Neznamov employed the term "engagement" (*srazhenie*) to define the scale of combat above battle and operation (*operatsiya*) to describe the linking together of maneuver and combat into a series of "individual bounds of the attacker forward and the defender backward."[69] The available technical means of control and communication were not, however, equal to the demands of time and space which the new weapons imposed.

Conscious of this transformation of the battlefield, Svechin took a position that would distance his approach from that of Tukhachevsky:

> We examine modern war, with all its possibilities and have not sought to fit our theory to red, Soviet strategic doctrine. To foresee the circumstance of the war in which the USSR might become involved is especially difficult, and one must approach any limitations on the general study of war with extreme caution. For each war one must select a specific line of strategic conduct with its own demands and avoid any stereotyped answers [*shablon*]. Each war is a special case, demanding assumptions [based] on their own logic. . . . Narrow doctrine can confuse us. War is a two-sided affair.[70]

Svechin claimed that his perspective was an objective one and noted that "our young critics" might accuse him of assuming "the pose of an American observer."[71] The most important tool in the development of strategic insights was history. "Strategy represents systematic reflection on military history."[72] Failure to utilize historical insights could only lead to disasters. "Isolation from the historical base is as dangerous for the strategist as it is for the politician." Svechin had in mind analytical history, as opposed to any superficial narrative, and this involved reflections on events, especially when causes did not yield the expected outcomes. For example, military theorists had assumed, on the basis of past experience, that railroads aided the offense, but the First World War had shown that it had actually strengthened the defense.[73]

Turning to the Soviet-Polish War, Svechin used that campaign to address the problem of balancing political ends against means and set political objectives in keeping with the military means at hand.

> To meet this condition the political figure [*politik*] must have a correct understanding about the relationship of his own forces to the enemy's—political figures brings [*sic*] insights from history and politics and statistics of the opposing states to this process. The final formulation of the goal will be done by the politician only after a corresponding exchange of opinions with the strategist. It must aid strategy and not make more difficult strategic decisions.

He expressly rejected Tukhachevsky's argument that Soviet strategy had been politically sound:

> In July-August 1920 Soviet policy put forward as a political goal for the continuation of the war with Poland the taking of Warsaw. This goal did not correspond to the relations of forces of the Red and Polish Armies and therefore raising it was a mistake. Even the formulation was wrong: Warsaw lies on the left bank of the Vistula River, the most important water boundary in the theater of war. The taking of Warsaw in the military sense involved the seizure of the left bank of the Vistula in at least 3-4 crossings on a front of several hundred versts. The seizure of Warsaw demands that an army occupying the entire right bank of the Vistula, undertake new, extraordinary operations, the complexity of which can be compared with the task of forcing the Rhine by the French before the Treaty of Versailles. If politics rendered an accurate report then the strategic goal could have been better set and the Red Army not placed in so impossible a position.[74]

Svechin suggested that a truly effective balance between political ends and military means in such a case could only be achieved by an "integral great captain," by which he meant a collective civil and military leadership adapted to the prosecution of war in all its dimensions. The "integral great captain" (*integral'nyi polkovodets*) could not solely be the commander in chief in the field. "The "strategist—the commander in chief—represents only part of the leadership of war."[75] This collective leadership prepared the country for war diplomatically, militarily, economically, socially, and politically. It conducted the struggle for the technological initiative, prepared the internal front against subversion, and mobilized the population to sacrifice for a just cause.

Svechin's emphasis upon the strategic advantages that accrued to the state fighting a politically defensive war was a critical element of his argument for a strategy of attrition. A national leadership that failed to address these issues before resorting to war, as had been the case of the Tsarist government in 1904 and 1914, was guilty of adventurism and could expect to taste the bitter fruits of defeat and internal unrest.[76] In 1920 the Soviet campaign against Poland had violated these strategic principles in a vain attempt to resurrect the strategy of annihilation:

> The Napoleonic knockout punch for winning a war with a single blow had been resurrected and colored red. However, on the way to the Vistula, the Red Army, just like the German Army on the way to the Marne, was unable to achieve extraordinary victories, on the final phase of the offensive the influence of geographic factors began to make itself felt.... In the Danzig corridor the Red Army sought to cut off not just the logistics of the Polish Army but the most important artery of the Polish state. The Red armies, ignoring the material forces of the Poles on the frontline, joined battle with the Versailles treaty. This is already mystification particularly under the conditions of annihilation.[77]

In a review of Russian military reform and restructuring in the decade before the First World War, Svechin concluded that a strategy of attrition (*izmor*) as opposed to annihilation (*sokrushenie*) fit Russia's strategic situation. In the end, the very nature of the Russian state made it suited for a strategy of a protracted war of attrition, not a war of annihilation:

> This process took place unnoticed even by the very leaders of reform in the army. . . . Russia's force for annihilation had not increased during those 14 years [1900-1914]. In this direction, which the evolution of Russian military power took, the single correct decision would be not an immediate campaign against Berlin, but a struggle for a further stage of deployment on the front Danzig-Peremyshl'.[78]

Svechin depicted a strategy of annihilation as a high-risk option in modern war. By structuring one's forces for a single initial blow, one ran grave risks if the enemy were not completely defeated, precisely because the choices that increased the chances of initial success undermined the ability to engage in protracted war. A force structured for a strategy of annihilation would trade off mobilization potential for professional readiness. Moreover, by pressing on to success in the initial operation one gambled that victory in a general operation over the first echelon of enemy forces would decide the course of the war. In fact, any such general operation carried with it a cloud of unknowns and the unfolding of the operation was likely to be like a "kaleidoscope show." Objective conditions tended to make a strategy of annihilation a high-risk venture. Svechin cited several factors contributing to this situation. The first was the absence of "deep battle capabilities" (*nedal'noboynost'*) which limited the possibilities of sustained operations over any depth, since the attacker had to wait for his repair of rail lines to reestablish and sustain the logistical tie between front and rear, necessary to the conduct of further operations.

Finally, modern states had the ability to generate additional strategic echelons to re-create a defensive line. Svechin's discussion of the term "mobilization" transformed its meaning from the initial deployment of armies to the total mobilization of nation, state, and economy. These new circumstances led to operational pauses and the emergence of positional warfare. A strategy of annihilation did have its utility, but that was against small states or large ones "in a state of political collapse."[79] In other cases, however, a strategy of annihilation involved unacceptable risk because operational and tactical considerations took precedence over the strategic.[80] In this regard Svechin's discussion of operational art stressed the idea of successive operations with limited goals to achieve strategic results. Pressing operational successes beyond culmination risked disaster. The Red Army's offensive in July 1920 had adequate combat power to push the Poles "back to the Neman and the Bug but not the Vistula."[81] These operational considerations were at the heart of his advocacy of a strategy of attrition in modern warfare. Attrition was not synonymous with positional warfare, although there was the risk of operations developing into positional struggles because of the necessary pauses imposed by operational exhaustion of the attacker. Nor was it an easy strategy to execute. Destruction, with its clarity of purpose and emphasis upon the overwhelming concentration of combat power to achieve immediate decision, was much more straight-forward.

Attrition demanded much greater acumen in the planning of successive operations to link them into a theater campaign. Such planning was at the heart of operational maneuver. On the Belorussian-Polish front in a future war Svechin anticipated significant opportunities for maneuver because of the anticipated lower densities of forces there. The key to success was effective strategic leadership linked to operational command and control, making possible the operational regrouping of forces to conduct successive operations, relying on economy of force and the determination of the subsequent strategic direction for the conduct of subsequent operations.[82] In the second edition of *Strategy*, published in 1927, Svechin was even more explicit—evolution of military art since the time of Moltke "has been running from destruction toward attrition."[83] This argument was further developed by Svechin's contemporary Vladimir Melikov in his study of the Marne, Vistula, and Smyrna operations. In his conclusion, Melikov stressed the high risks of a strategy of annihilation (*strategiya va bank*) during the initial period of war and the associated emphasis on operational Cannaes, which he labeled "Cannomania."[84] Despite such support, Svechin's critics, who were numerous within the Red Army, accused him of underestimating the potential of a strategy of annihilation. That debate surfaced in the continuing effort to prepare the Soviet Union for future war.

OPERATIONAL ART AND FUTURE WAR

Within a year of Svechin's invocation of the term, operational art became the subject taught in the new Chair on the Conduct of Operations within the Department of Strategy at the Military Academy of the RKKA. Svechin's conceptualization of operational art coincided with Mikhail Frunze's appointment as chief of staff of the RKKA and chief of the Military Academy. At Frunze's initiative, the Chair of Army Operations was established in the Department of Strategy at the Academy of the RKKA in 1924 but did not survive for long.[85] The Chair disappeared within the year, only to reappear as a department in the Frunze Military Academy in 1931. Even so, the very existence of this new category within the Soviet military had a profound impact on Soviet military art, military doctrine, and the concept of future war.[86]

Frunze played a leading role in reinvigorating the Academy's higher military-academic courses (*VVAK*) for senior Red Army commanders, which focused on the further education of brigade and higher commanders.[87] Frunze's commitment to this program brought more attention to the Chair and its further development. He emphasized the need to change the content of the course on the conduct of operations by shifting from general observations to working out the practical details and techniques concerning the art of command in conducting operations.[88] Over the next several years this led to the development of a program of operational war-gaming in which students were expected to do the calculations and estimates necessary to prepare for an army operation. This "applied" approach to training future commanders and staff officers was a major break with past Russian tradition and placed primary stress upon finding means in the educational process of unifying theory and practice. The leaders in the development of operational war-gaming at the Academy were V. K. Triandafillov, K. Berends, and N. Varfolomeev.[89] The summer campaign of 1920 served as both a model and a case study for such operational gaming, since it embraced a major operational axis in a war against one of the most probable future opponents of the Soviet state.

Tukhachevsky, who served as deputy chief of staff to Frunze in 1924-1925, took over the Chair of Strategy. In this capacity he worked closely and effectively with Svechin, even though their strategic concepts were radically different. Each man had a set of supporters within the Academy. It seems that as long as Frunze lived this debate could go forward within the Academy in spirited but hostile fashion. Shortly after his death this situation changed sharply. In early 1926 at a special conference held to debate the merits of strategies of attrition (*izmor*) and annhilation (*sokrushenie*) faculty members from the Military Academy and officers of the Main Staff of the RKKA took opposing sides.[90]

Svechin's argument for a national strategy based upon attrition had its roots in his own vision of Russian society and the historical experience of the First World War. His fellow professor and colleague A. Verkhovsky, in defending an attrition strategy, enraged the offensive-minded young Red commanders when he asserted that it might be better in the initial period of a future Polish-Soviet war "to give up Minsk and Kiev than to take Bialystok and Brest." To those who identified Marxism-Leninism with a strictly offensive style of war, such retreats were quite unthinkable.[91]

As the Red Army's leading author on tactics, Verkhovsky championed preparing the army for battle with a concrete enemy in specific circumstances. The features which marked this "new school" of tactics from the old were, a), the characteristics of one's own weapons; b) the influence of class and national conflict within which a future war would be fought; c) the quantity of troops available to the enemy and the Red Army, and the size of the theater, density of forces, and depth of deployments; d) how the opponents will act "not with our weapons but with his and according to his own regulations which are in keeping with his weapons and his troops"; e) the decisive influence of locality in the sense of both theater of war and within the confines of the field of battle; and f) finally, application of the closest and most intense scrutiny in calculating the influence of the element of time on the forms of struggle and on the degree of its organization not only on the enemy side but also on the Soviet side.[92] All these points, while touching upon strategic topics in one way or another, addressed operational issues. Density and depth of forces as expressed as number of troops and guns on a given front could be reduced to calculations of density of forces per kilometer of front. "Without calculations all these forms lack content. Furthermore, it is very important to know the density of forces in a given front at which the saturation point is reached in those cases when we wish to set the form of a march-maneuver in a future war."[93] Verkhovsky's vocabulary at once pointed toward the past in its emphasis on march-maneuver and toward the future in its stress on operational correlations of forces to create the conditions for deep battle.

Thanks to the intervention of the newly appointed deputy chief of staff of the RKKA, M. N. Tukhachevsky, operational art became the domain of N. Varfolomeev, the deputy head of the Department of Strategy during the same period. Varfolomeev noted the fact that objective changes in the nature of warfare associated with the appearance of million-man armies and technological innovations had recast the face of battle, increased its spacial and temporal dimensions, broken down the conventional forms of combined arms, forced a rethinking of problems of command and control, and laid the foundation for the emergence of the operation as the bridge between strategy and tactics. Tactics became the conduct of battle/combat (*boi*). The engagement (*srazhenie*) that in the Napoleonic era had been conducted as a series of combats on a single battlefield under the observation of the commander, would now take place over a much broader front and at much greater depths, well beyond the ability of any single commander to exercise direct control. In this manner the operation emerged as the bridge to strategy. Varfolomeev described the modern operation as "the totality of maneuvers and battles in a given sector of a theater of military actions [TVD] which are directed toward the achievement of a common objective, which has been set as final in a given period of the campaign. The conduct of an operation is not a matter of tactics. It has become the lot of operational art."[94] Warsaw figured prominently in Varfolomeev's own further research on the problem of pursuit and exhaustion and the role of shock armies in breakthrough operations.[95]

RECONCILIATION AND THE MECHANIZATION OF WAR

The synthesis of the positions laid out by Tukhachevsky and Svechin in the mid-1920s found its public expression in V. K. Triandafillov's *The Nature of the Operations of Modern Armies*, published in 1929.[96] Triandafillov emerged as one of the most important advocates of an operational art adapted to the realities of a future war, fought on the basis of a continuation of the New Economic Policy (NEP). The logistical parameters of deep, successive operations to a great extent depended upon the visions of the Soviet Union as a political economy and the nature of the external threat. In the hands of Svechin and those like him who emphasized the need to prepare for a long war, the maintenance of the workers' and peasants' alliance became the central reality of the Soviet Union's domestic mobilization base. Such a view assumed that Lenin's NEP, with its emphasis upon agriculture's recovery, would be the long-term policy of the USSR. At the same time, such authors cast the nature of the external threat in terms of the states immediately bordering the USSR. Such authors could not ignore postwar developments in military technology, but they concluded that Europe was, in fact, divided into two parts, two military-technical systems. The West was industrial, and the potential for a mechanization of warfare was there to be seen. Eastern Europe, which included the USSR, was dominated by a peasant economy and a "peasant rear" (*krest'ianskiy tyl*).[97]

Triandafillov had served in the Tsarist army during the First World War, took an active part in the revolutionary politics within the army in 1917, and joined the Red Army in 1918, where he commanded a battalion, regiment, and brigade. He fought on the Ural Front and on the Southern and Southwestern Fronts against Denikin and Wrangel. Joining the party in 1919, he was a natural choice for education as a Red general staff officer *(genshtabist),* and was posted to the Academy in the same year. During his four years with the Academy he divided his time between theory and practice. As a brigade commander with the Fifty-first Rifle Division, one of the best in the Red Army, he took an active part in Frunze's successful offensive at Perekop Isthmus against Wrangel. At the same time, Triandafillov began writing military analyses of operations in the Civil War as his part in the activities of the Academy's Military Scientific Society. These included essays on Southern Front's offensive against Denikin and the Perekop offensive.[98] He also took part in the suppression of the Tambov insurrection in 1921, where he served under Tukhachevsky. Like Varfolomeev, Triandafillov also wrote on the Soviet-Polish War. However, whereas Varfolomeev had concentrated on the problem of pursuit during a general offensive, Triandafillov used a small-scale action from the final phase of the war, Twenty-seventh Omsk Rifle Division's action against the Polish Fifteenth Poznan Division near Volkovysk in mid-September 1920, to address the problem of troop control at the tactical level as it contributed to a force's achievement of surprise and its vulnerability to unexpected combat developments.[99] Following Triandafillov's graduation from the Military Academy in 1923, Frunze chose his former subordinate to join the Main Staff of the RKKA, where he took over as chief of the Operations Section in 1924. From there he moved on to command a rifle corps and then returned to Moscow as deputy chief of staff for RKKA in 1928.

Charged with putting operational art into practice, Triandafillov authored what became the chief work on the nature of the operations of modern armies. The work laid out in detail the military context of the theory of successive deep operations. Triandafillov called attention to the process of technological development which was making possible the "mechanization" of warfare, but noted its limited impact upon the economically backward regions of Eastern Europe with their peasant rear. New automatic weapons, armor, aviation, and gas would affect such a war but would not become decisive. He also treated the problem of manpower mobilization and the reality of mass war quickly becoming a war of conscripts and reservists. This brought him to the problem of addressing the means of achieving breakthrough and sustaining pursuit in successive deep operations. Here he drew upon Frunze's use of shock armies against Wrangel for the breakthrough and the employment of echeloned strategic cavalry forces to facilitate exploitation and pursuit.

Much of the success in such operations turned upon two related problems: the organization of an effective command and control system to coordinate the operations of several fronts and the establishment of realistic logistical norms in keeping with the geographic-economic realities of the theater of military action.[100]

As deputy chief of staff to the RKKA Triandafillov's views reflected some basic assumptions regarding the sort of war the Red Army would fight in the future. The Field Regulations of 1929, in their treatment of the offensive, touched on many of the same themes developed by Triandafillov in greater depth.[101] While the new regulations did provide for successive deep operations based upon a combined arms offensive, the armies described by Triandafillov and the regulations were modernized versions of the Red Army from the Civil War. This vision was in keeping with what Boris M. Shaposhnikov had described as the political-military context of Soviet strategy in his classical work, *The Brain of the Army.*[102]

Triandafillov died in an airplane crash in 1931 before he had a chance to complete a new and revised edition of his book. The outline for this revision, which was published in the posthumous editions of his book, do contain some clues as to the major changes which he envisioned. First, in keeping with the new party line on the external threat, Triandafillov addressed both the crisis of capitalism and the increased risk of direct attack upon the USSR by one or more major capitalist powers. Second, he began to address the problem of employing massed armor in the offensive. The first Five-Year Plan had promised to industrialize the USSR, and now it was possible to put the USSR within the ranks of the modern Western European states and the United States. Third, Triandafillov specifically turned his attention to the problem of mechanized combined arms in the conduct of deep operations. The outline is at best a sketch without details. Soviet officers have been willing to say that these few remarks anticipate the mechanization of successive deep operations as presented in the 1936 Field Regulations.[103]

There were other advocates of operational art that argued technological developments and the nature of the external threat made it absolutely essential to carry out a total mechanization of the Red Army and Soviet rear. One of the leading proponents of such views was M. N. Tukhachevsky, who was Triandafillov's immediate boss as chief of the RKKA Staff from 1925 to 1928. Tukhachevsky argued that what was required to make the new operational art into a sound strategic posture was nothing less than "complete militarization" of the national economy to provide the new instruments of mechanized warfare. Committed to an operational art which would end in the total destruction of the enemy, Tukhachevsky crossed pens with Svechin, whom he accused of being an advocate of attrition.[104] According to G. S. Isserson, one of his closest collaborators in the 1930s, during the war scare of 1927, when the party leadership feared conflict with Great Britain, Tukhachevsky came forward with a master plan for the mechanization of the Red Army in December, only to have it turned down by the party leadership under Stalin.[105]

This study, which was done by a group of researchers from the Fourth Directorate of the Staff of the RKKA (Main Intelligence Directorate) under Tukhachevsky's direction, was circulated in 1928 in a limited edition to central administrative organs and to the military districts. Its topic was "future war", that is, a future conflict for the USSR, taking into account the general political situation, the human resources available to the USSR and its probable adversaries, economic factors affecting supply and logistics and including the economic bases of war potential, technological factors and the influence of weapons modernization and innovation on the nature of future war, political factors and an assessment of class, agrarian, and national conflicts within probable opponents, and operational and organizational problems affecting the conduct of war. As Jan Berzin, the head of the Fourth Directorate, pointed out in his introduction, the task of adapting a state's military system to the needs of future war and preparing a state and military for future war was common to all states. Berzin noted the mistakes made by the European powers in their preparations prior to the First World War and identified their basic mistake as underestimating the changed conditions brought about by the development and hegemony of imperialism. By this Berzin meant those characteristics associated with total war, the "monstrous material scale of the war, the unprecedented intensity of the struggle, the colossal shocks in the areas of economic and political life."[106] The armies and states of Europe, including their general staffs, were not prepared for the war that they faced. Berzin discussed the origins of this project and called attention to the fact that in 1926 Tukhachevsky, as chief of the RKKA Staff and in response to the Fourteenth Congress of the Communist Party, had ordered a study of future war in keeping with the Party's directives on the industrialization of the country.[107]

An important aspect of this study is the primary focus of the threat, which is seen as Poland and Rumania, along with the other successor states along the Soviet Union's western border. The study addresses their existing military systems, mobilization potential, and industrial base, and examines the possibilities of external military assistance to these potential adversaries from the leading imperial powers. It makes a very powerful case for the militarization of the Soviet economy to meet these threats and examines a wide range of operational issues. The final chapter of the study was a ringing endorsement of the industrialization of the national economy to meet the needs of the military.[108] In the end, Tukhachevsky's study came down on the side of preparing for a future war on the basis of "total war." In operational terms it looked to improving the ability of the Red Army to conduct deep battle, through increasing the "far-battle character (*dal'noboinost'*) of contemporary operations."[109] The study went on to embrace the concept of logistical constraints on such operations, noting the limits that rail throughput capacity placed on the support of large-scale offensive operations in terms of the effective distance forces might advance before exhaustion set in. Motorization might lessen but could not eliminate this problem. Thus, operational pauses and the regrouping of forces had become a necessity.[110]

The development of deep battle capabilities went hand in hand with the planning for successive operations. On this issue, Tukhachevsky's study ran into a distinct problem. On the one hand, the Red Army had to prepare for the conduct of decisive operations in the initial period of war, but on the other hand, the Red Army lacked the means to achieve a rapid and decisive victory. Civil War operations, including the "campaign beyond the Vistula," were irrelevant and could not be compared with those of the First World War.

> Thus, just as the experience of the [First] World War so also the experience of 1920 shows that: a) one should not build on horse transport the logistical base of modern troops, even if very few in number, at a great distance; b) a sufficiently rapid re-establishment of RR [Rail Roads] behind advancing troops and the structure of their rear remains an unresolved task for contemporary technology. Therefore, the operational capabilities of contemporary armies still remain limited.[111]

The point became one of conducting each operation so that it would be decisive within its own depth. Within such bounds the object was to bring about the destruction of the opposing enemy forces throughout the depth of their deployments by means of breakthroughs and encirclements. On this point the study cited no less an authority than J. F. C. Fuller. The architect of Plan 1919 and the prime promoter of the mechanization of the British Army had written in 1926 of the possibility of using new technology in deep battle:

> At present aviation can attack the enemy rear; tanks can break through the front and attack the rear; armoured cars can turn his flank and once again attack his rear, i.e., mount attacks against the most sensitive part of the force, in his stomach. The attack of the rear at the present time is quite possible and in my opinion has become one of the most important tactical operations in war.[112]

These partial destructions could not prevent a large and economically developed state from redeploying forces to meet the threat and from mobilizing additional resources. However, the combination of such operations was the most likely road to decisive victory.[113] At the same time the study admitted that the threat of exhaustion and positional warfare could not be precluded. In that case, the Soviet Union had to prepare for protracted war and mobilize its entire economy and society.[114] The key to overcoming the threat of positional warfare was the mechanization of the armed forces to assist in breakthrough and exploitation:

> The density of the front in our theater of military actions has approached the density of the Russo-Austro-German Front of the beginning of the 1914-1917 war. Therefore from the best operational preparation of the Red Army for future war *the center of gravity of the study of the experience of past wars must be concentrated on the maneuver period of the 1914-1917 war and not on the operations of the Civil War. In the study of the Civil War for the future it is necessary to take, primarily, the political factors of the war.*[115]

A much more narrow critique of attrition strategy built on Svechin's own observation that in the initial period of war the attacker, the side adapted to decisive initial operations, could impose its style of warfare upon the defender. Viktor Novitsky noted that a strategy based upon attrition stood on totally different principles than one based upon annihilation. Annihilation required the ability to conduct large-scale, immediate, decisive, lightning operations. In place of mobilizing the civilian economy for war, an annihilation strategy required an in-place war industry which would in peacetime provide all the weapons and materiel necessary to conduct decisive operations. Svechin had assumed that the side which adopted a strategy of annihilation would be able to impose his war upon the other side by seizing the initiative and mounting initial offensive operations. Counting upon victory in a short war, the side adopting a strategy of annihilation could avoid a host of difficult peacetime sacrifices necessary to create a unity of front and rear in a protracted war. However, failure in those initial operations would expose the adventurism at the heart of such a policy, by underscoring the disconnection between military strategy and political-economic preparations. Novitsky reformulated Svechin's assumption about the initiative always going to the side following a strategy of annihilation by focusing upon the problem of the struggle for mobilization and deployment. He asserted that technological innovations were making "deep operations" potentially decisive in the initial period of war. In the age of airpower, he emphasized the possibility of a covering force army conducting initial operations so as to disrupt enemy mobilization and deployment and, thereby, to win the "struggle for the nature of future war."[116] Novitsky's work on this aspect of future war contributed to the development of a specific line of Soviet military writings devoted to the nature, form, content, and law-governed patterns (*zakonomernosti*) of the development of the "initial period of war."[117]

TWO VIEWS RECONSIDERED

Just as Stalin was consolidating his power and initiating the revolution from above, the debate about two Warsaws was resolved. Operational art had become an accepted category within military art. It remained to be seen what the relationship would be between operational art and strategy, and in that regard the Civil War still loomed large in defining the threat and formulating the political objectives to which military power would be applied. On that issue it was Stalin who now had decisive influence.

In 1930 Tukhachevsky's views won favor, as Stalin had begun to associate the Depression with a rising threat of war to the Soviet Union. This threat the party leadership openly used to justify the brutal processes of industrialization and forced collectivization by now linking them with an improvement in the level of national defense. In 1931 Stalin employed a basic calculus to justify the drive for modernization, in which he linked backwardness and defeat: "Those who fall behind, get beaten."[118]

During the intervening two years Tukhachevsky had left the RKKA Staff to take over as commander of Leningrad Military District, where he conducted a number of experiments relating to mechanization. These experiments came at a time when motorization and mechanization had emerged in Western Europe as alternative solutions to the problem of integrating the internal combustion engine into the armed forces. The former implied grafting automobile transport onto existing combat arms, while the latter called for the creation of "self-propelled combat means" with an emphasis upon armor, especially tanks, armored cars, and self-propelled artillery.[119] In his comments on the training exercises of the troops of the Leningrad Military District Tukhachevsky emphasized the need to increase their mobility as a combined arms force which could engage in a multi-echeloned offensive. His interest in the development of tank, aviation, and airborne forces during this period marked him as an advocate of mechanization.[120]

At the Sixteenth Party Congress and the Ninth Congress of the Komsomol in 1930-1931 K. E. Voroshilov, the commissar of war and Stalin's closest collaborator, spoke out regarding the mechanization of warfare as bringing about a qualitative change in the nature of future wars. But in Voroshilov's case mechanization would in the future bring about the possibility of a short, bloodless war, carried quickly onto the territory of the attacking enemy.[121] Such views emerged at a time when it appeared that world capitalism had gone back into a profound political-economic crisis which was creating greater instability and increased risks of war. This in turn, it was feared, had created the bases for the formation of a broad anti-Soviet alliance, which threatened war on every frontier. At home the strains of the first Five Year Plan were also underscoring the possibilities of an alliance between the external threat and the so-called internal enemy, the forces of counter-revolution.

Stalin had already put that face on the so-called Shakhty Affair at the April plenum of the Central Committee of the party in 1928. His "facts" were that there was an "economic counter-revolution," led by bourgeois technical specialists (*spetsy*), and funded by capitalist organizations in the West to sabotage the Soviet coal industry. Stalin linked this "economic intervention of West-European, anti-Soviet capitalist organizations" with the earlier military-political intervention of the Civil War. In both cases the appropriate answer was to liquidate the threat, and in both cases the threat came from class enemies, in this case bourgeois specialists, who put their talents at the service of the encircling capitalist powers. Stalin warned: "We have internal enemies. We have external enemies. Comrades, we cannot forget about this for even one minute." From *spetsy* to kulaks, to wreckers within the very highest reaches of the party itself, that was the terrible logic of Stalin's campaign against wreckers and enemies of the people, a term applied to both Tukhachevsky and Svechin.[122]

In 1930 Tukhachevsky presented his own powerful arguments for a mass, mechanized army as the means to execute the new operational art. He used many forums to present this argument. One was the foreword to the Russian translation of Hans Delbrück's *Geschichte der Kriegskunst im Rahmen der politischen Geschichte*, which provided a forum in which to attack Svechin's concept of attrition as the appropriate strategy for the USSR.[123] This work was conspicuous for the tenor of the political-ideological assault mounted by Tukhachevsky against the old *genshtabist*. In a time of heightened suspicions toward all specialists as wreckers, Tukhachevsky called his colleague an "idealist" in Marxist dress.

Worse attacks followed within the confines of the Section for the Study of the Problems of War in the Communist Academy, which was organized in 1929 as part of an effort to infuse Marxism-Leninism into military science.[124] Within the Section, as within the Communist Academy, the notion of a struggle between an old bourgeois past and a young, dynamic Communist future were given free reign. The debates over "unified military doctrine" of 1921-1922 were recalled but now within the context of a struggle over the issue of where the center for the study of military problems in the USSR was to be. The leaders of the Section were promoting it as a rival to the Military Academy and looked to enhance their position through party ties and by building "strong ties with the Institute of Red Professorship and those young Marxist-Leninist forces which now move our Bolshevik science."[125] Tukhachevsky, armed with the appropriate citations from

Lenin, Stalin, and Voroshilov, attacked Professors Svechin and Verkhovsky. He described their writings as infested with bourgeois ideology. In Svechin's case the fault was that he did not believe in the possibility of decisive operations but defended the idea of limited war. Verkhovsky was charged with favoring a professional army at the expense of mass. Tukhachevsky spoke positively of Triandafillov's book, which had critiqued Verkhovsky's concept of a cadre-mechanized forces, but noted some shortcomings.[126] His line of criticism fit that offered in a review of Triandafillov's book published in the spring of 1930. The reviewer took the author to task for talking of a peasant rear without noting the possibility of industrializing that rear area. That industrialization, the reviewer pointed out, would make it possible to speed up the massing of forces and their maneuver, creating opportunities for decisive operations, if the political and revolutionary possibilities were exploited.[127] As we have noted above, Triandafillov was himself responding to those new possibilities when he died in 1931.

That same year Tukhachevsky became deputy commissar of military and naval affairs, a member of the Revvoensovet, and director of armaments for the RKKA. Over the next six years he directed the mechanization of the Red Army, laying the foundations for the creation of a mass, mechanized force designed to conduct successive deep operations in a war of annihilation.

Soviet military historians tended to emphasize the infusion of Marxism-Leninism into Soviet military science as the dominant and decisive theme in the development of Soviet military theory in the entire interwar period and downplay the significance of Stalinization or the debate among contending groups in the early 1930s. Stalinization in its military manifestation was closely connected with both the substance and style of Stalin's "revolution from above" as it developed during the first Five-Year Plan. With regard to substance it assumed the form of a major rearming of the Red Army thanks to the fruits of forced industrialization, making it possible for the Soviet military to leap from an infantry-cavalry army based upon a peasant rear (*krest'ianskiy tyl*)—to use Triandafillov's term—during the period of the New Economic Policy (NEP), 1921-1928, to a mechanized force based upon an industrial rear by the end of the second Five-Year Plan. In style Stalinization of the Soviet military meant the revival of the class antagonisms and hostilities of War Communism but now conducted ruthlessly against potential class enemies. While drawing upon the examples of the Red Terror from the Civil War for inspiration, the new repression was shaped by an ideological dogmatism, and, unlike War Communism, when the cause was the destruction of the bastions of hostile classes, the Stalinist campaigns against wreckers, traitors, and enemies of the people now sought to create an invulnerable Soviet bastion. Such a style was not just a function of the whim of a single man or even a narrow circle surrounding him but a cultural and psychological legacy of War Communism which shaped the worldview and values of much of the party and Soviet apparatus. While this was not the only legacy of Bolshevism, as Stephen Cohen has asserted, it proved a powerful and compelling one.[128]

The Stalinist industrialization did make the USSR into a major industrial power with the capacity to mechanize its armed forces to an extent undreamed of by Triandafillov. During that same period the nature of the military threat confronting the USSR became more complex and serious. To his credit Tukhachevsky never fell into the trap of assuming that mechanization would negate mass war. He was an informed critic of "blitzkrieg theory," and his criticism of the works of Fuller, Liddell Hart and others deserves serious attention. They contain a good clue about the emerging Soviet way of war. In 1931 he wrote as follows regarding the professional mechanized army:

> Let's imagine a war between Great Britain and the USA, a war, for example, which breaks out along the Canadian border. Both armies are mechanized, but the English have, let's say, [J.F.C.] Fuller's cadres of 18 divisions, and the U.S. Army has 180 divisions. The first has 5,000 tanks and 3,000 aircraft, but the second has 50,000 tanks and 30,000 planes. The small English Army would be simply crushed. Is it not already clear that talk about small, but mobile, mechanized armies in major wars is a cock-and-bull story? Only frivolous people can take them seriously.[129]

Thus, in Tukhachevsky's work, Soviet military theory, building upon the work of the Tsarist general staff and the combat experience of four industrial wars, namely the Russo-Turkish, Russo-Japanese, the First World War, and the Civil War, focused on the mechanization of the mass army as the means to conduct decisive operations in a total war. For Tukhachevsky independent tank and mechanized formations were the keystone to such deep operations. The "long-range tanks," which would make up such mobile groups, had to be high-speed, rugged, reliable, and most of all armed with a heavy cannon to fight and defeat enemy tanks.[130] Tukhachevsky's collaborator, G. Isserson, provided the intellectual synthesis for this development in operational art, which he depicted as the latest stage in the evolution of strategy: from the Napoleonic strategy of the "single point," to Moltke's strategy of the extended line and the crisis of linear warfare of the First World War, and its negation, "deep strategy," to the use of new means of deep battle to conduct deep operations to bring about the annihilation of an opposing force throughout the depth of its deployments.[131] New technical means deployed in shock armies had made possible breakthroughs into the tactical depths of the enemy deployments, but these formations could not carry the struggle into operational depths. A second echelon, based on new motorized-mechanized formations and cavalry and supported by airborne assaults in the enemy's operational rear, would exploit such breakthroughs to their operational depths and now offered the possibility of destroying an entire front.[132]

Ironically, the deal that Tukhachevsky and his allies struck to get Stalin's support for the realization of the material and technical means to execute their concept of deep operations proved costly not only to their opponents but to themselves, the Red Army and Soviet society. Both Svechin and Tukhachevsky died at the hands of Stalin's terror and became nonpersons. In 1941 a massive

Red Army, equipped with thousands of obsolescent tanks and aircraft and led by junior officers promoted rapidly to fill the vacancies left by Stalin's blood-purge of the army, nearly collapsed in the face of the Wehrmacht's initial onslaught. Then Svechin's mass army, trading men and space for time, fought a war of attrition until new mechanized forces could be created and their leadership trained for deep operations by the terrible test of battle.

NOTES

1. The views expressed in this chapter are those of the author and should not be construed to represent the views of the Department of the Army or the U.S. Department of Defense.
2. U.S. Department of the Army, *Operations FM 100-5* (Baltimore, 1982), 2-3.
3. James J. Schneider, "Theoretical Implications of Operational Art," in Clayton R. Newell and Michael D. Krause, eds., *On Operational Art* (Washington, DC, 1995), 26-29.
4. Condoleezza Rice, "The Making of Soviet Strategy," in Peter Paret, ed., *Makers of Modern Strategy from Machiavelli to the Nuclear Age*, 2nd edition, (Princeton, NJ, 1986), 648-76.
5. Stalin had his own political reasons for suppressing discussion of the causes for the failure in the blow against Warsaw. For a sycophantic account, see, K. E. Voroshilov, *Stalin i vooruzhennye sily SSSR* (Moscow, 1951). On Stalin's cult see Stephen F. Cohen, "Bolshevism and Stalinism," in Robert C. Tucker, ed., *Stalinism: Essays in Historical Interpretation* (New York, 1977), 3-27.
6. Richard Simpkin, *Deep Battle: The Brainchild of Marshal Tukhachevskii* (London, 1987), ix.
7. Ibid., 249-70.
8. Jacob W. Kipp, "Lenin and Clausewitz: The Militarization of Marxism," *Military Affairs* XLIX, no. 4 (December 1985), 184-91.
9. Jozef Pilsudski, *Rok 1920*, 5th edition. (London, 1941); Jozef Pilsudski, *Year 1920* (London, 1972). Cf. J. F. C. Fuller, *Decisive Battles: Their Influence Upon History and Civilization*, (New York, 1940), 973.
10. Norman Davies, *White Eagle, Red Star: The Polish-Soviet War, 1919-1920* (New York, 1972), 130-33.
11. M. Tukhachevsky, *Pokhod za Vislu* (Smolensk, 1923), 1.
12. Anthony John Trythall, *"Boney" Fuller: Soldier, Strategist, and Writer, 1878-1966* (New Brunswick, NJ, 1977).
13. M. Tukhachevskiy, "Voyna klopov," *Revolyutsiya i voyna*, no. 22, (1923), 189.
14. Shimon Naveh, "Michail Nikolayevich Tukhachevsky," in Harold Shukman, ed., *Stalin's Generals* (London, 1993), 262-63.
15. Lev Nikulin, *Tukhachevskiy: Biograficheskiy ocherk* (Moscow, 1964), 30-40.
16. A. S. Bubnov, *O krasnoy armii* (Moscow, 1968), 216.
17. D. A. Kovalenko, *Oboronnaya promyshlennost' Sovetskoy Rossii v 1918-1920 gg.* (Moscow, 1970).
18. M. N. Tukhachevsky, "Pokhod za Vislu," *Izbrannye proizvedeniya*, I, (Moscow, 1964), 126-27.

19. R. Tsifer, "Zametki o voynena malokul'turnykh teatrakh i metode ee izucheniya," *Voyna i revolyutsiya*, no. 11 (1928), 132-40.

20. Lev Nikulin, *Tukhachevskiy: Biograficheskiy ocherk* (Moscow, 1963), 161.

21. Tukhachevskiy, "Pokhod za Vislu," *Izbrannye proizvedeniya*, I, 142-43.

22. Tukhachevskiy, "Revolyutsiya izvne," *Revolyutsiya i voyna*, no. 3 (1920), 45-54.

23. Ibid., 46. See also A. A. Svechin, "Opasnye illyuzii," *Voyennaya mysl' i revolyutsiya*, no. 2 (1924), 49-50.

24. "Konnitsa v grazhdanskoy voyne," *Revolyutsiya i voyna*, no. 6 and 7 (1921), 36.

25. A. I. Soshnikov et al, *Sovetskaya kavaleriya: Voyenno-istoricheskiy ocherk* (Moscow, 1984), 3-24.

2626.. A chief provincial-administrative region.

27. M. Ryshman, *Reyd Mamontova, avgusta-sentiabrya 1919 g.* (Moscow, 1926), 16-29.

28. Ibid., 30-43.

29. K. Monigetti, *Sovmestnye deistviya konnitsy i vozdushnogo flota* (Moscow, 1928), 92-93.

30. Hoover Institution, Wrangel Papers, Box 33 (*delo* 146), Arkhiv Shtaba Glavkomand-ago V[ooruzhennykh] S[il] na yuge Rossii,"Svodki i doneseniya razvedyvatel'nykh punktov Glavkom-ago V. Silami na yuge Rossii za period Yanvar'-Noyabr' 1918 goda," Nachal'nik, Khar'kov Razved. Punkt, 9 Sentyabr' 1919 g., no. 132.

31. Soshnikov, *Sovetskaya kavaleriya*, 62-63.

32. I. Babel, *Konarmiya*, in, *Izbrannye proizvedeniya* (Moscow, 1966), 27-58.

33. Zotov, "Boi 1 konnoy armii v raione Rovno v iyune 1920 g.," *Voyna i revolyutsiya*, no. 2 (1929), 102-3; and U.S. Army, Attache Reports (Poland), no. 1095, "Operations of Budenny's Cavalry", (9 December 1920), 1-4.

34. Zotov, "Boi 1 konnoy armii v raione Rovno v iyune 1920 g.," *Voyna i revolyutsiya*, no. 2, (1929), 104-18. On other uses of strategic cavalry in the Civil War see M. I. Vladimirov et al, *M. V. Frunze: Voennaya i politicheskaya deyatel'nost'* (Moscow, 1984), 137-47; and V. V. Dushen'kin, *Vtoraya konnaya: Voyenno-istoricheskiy ocherk* (Moscow, 1968), 189-206.

35. Pilsudski, *Year 1920*, 83.

36. Nikulin, *Tukhachevskiy*, 119-22.

37. M. Weygand, *Mémoires: Mirages et Réalité* (Paris, 1927), 115.

38. Tukhachevsky, "Pokhod za Vislu," *Izbrannye proizvedeniya,* I, 134-52; and Pilsudski, *Year 1920*, 85-150.

39. Pilsudski, *Year 1920* 154-57.

40. Ibid., 151-208.

41. V. Triandafillov, "Vziamodeystvie mezhdu zapadnym i iugozapadnym frontami vo vremya letnego nastupleniya krasnoy armii na Vislu v 1920 g.," *Voyna i revolyutsiya*, no. 2, (1925), 26-27.

42. The extent of Soviet military studies on the Vistula operation of 1920 is recorded in a bibliography on the Soviet-Polish War prepared by the Military Section of the Communist Academy in 1930 which listed 257 titles, most of them Soviet books and articles. See "Bibliograficheskiy ukazatel' literatury po sovetsko-pol'skoy voyne 1920 g.," in *Kommunisticheskaya akademiya, Sektsiya po izucheniyu problem voyny, Zapiski* I, (1930), 219-31. The Stalinist version of events is summed up in I. Apanasenko's essay on the *Konarmia*. See I. Apanasenko, "Pervaya konnaya," *Voyenno-istoricheskiyzhurnal*, no. 4 (November 1939), 35-42.

43. Tukhachevsky, "Pokhod za Vislu," 167.

44. Kipp, "Lenin and Clausewitz," 184-91.

45. Fedyukhin, *Sovetskaya vlast' i burzhuaznye spetsialisty*, (Moscow, 1965), 77.

46. I. A. Korotkov, *Istoriya sovetskoy voyennoy mysli: Kratkiy ocherk 1917-iyun' 1941*, 32-34.

47. A. Verkhovskiy, "Evolyutsiya prepodavaniya taktiki v 1918-1928 gg.," *Voyna i revolyutsiya*, no. 11 (November 1928), 50-52. On Verkhovsky, see *Voyennyy entsiklopedicheskiy slovar'* (Moscow, 1983), 126. A similar characterization of the debate within the Red Army is described by A. Heroys and Léon Thevinin in A. Heroys and Léon Thevinin, *L'Armée Rouge et la Guerre Sociale* (Paris, 1931), 13-17.

48. A. A. Neznamov, *Osnovy sovremennoy strategii lektsii, chitannye A. A. Neznamovym na uskorennom kurse Akademii General'nogo Shtaba Raboche-Krest'yanskoy Krasnoy Armii* (Moscow, 1919), 30.

49. Ibid., 39-45.

50. B. Shaposhnikov, *Na Visle. K istorii kampanii 1920 goda* (Moscow, 1924), 17-20, 200-205.

51. N. Varfolomeev, "Strategiya v akademicheskoy postanovke," *Voyna i revolyutsiya*, no. 11 (November 1928), 84.

52. A. A. Svechin, *Strategiya*, 2nd edition (Moscow, 1927), 14 ff.

53. A. Svechin, *Predrazsudki i boevaya deistvitel'nost'* (St. Petersburg, 1907), 1.

54. A. A. Svechin, "Del'bryuk—istorik izmora i sokrusheniya," in A. A. Svechin, ed., *Strategiya v trudakh voyennykh klassikov*, II (Moscow, 1926), 233-35.

55. Ibid., 235.

56. A. Svechin, *Evolyutsiya voyennogo iskusstva*, II (Moscow, 1927-1928), 537 ff.

57. I. A. Korotkov, *Istoriya sovetskoy voyennoyi mysli*, 28.

58. A. Svechin, "Trudy Komissii po issledovaniyu i ispol'zovaniya opyta voyny 1914-1918 gg.," *Voyenno-istoricheskiy sbornik*, I, (1919), 3-8.

59. Ibid., 6.

60. Ibid., 9.

61. A. I. Reznichenko, ed., *Akademiya imeni M. V. Frunze: Istoriya Voyennoy ordena Lenina, Krasnoznamennoy ordena Suvorova Akademii* (Moscow, 1972), 40.

62. Jacob W. Kipp, "General-Major A. A. Svechin and Modern War," in Kent Lee, ed., *A. A. Svechin, Strategy* (Minneapolis, 1992), 23-60.

63. Ibid., vi.

64. A. A. Svechin and Romanovsky, *Russko-yaponskaya voyna, 1904-1095 gg.*, 337-81. See also *Voennaya entsiklopediya*, XIV (St. Petersburg, 1914), 474 ff.

65. A. N. Kuropatkin, *Deistviya otryadov generala Skobeleva v russkoturetskuyu voynu 1877-1878 godov*, II (St. Petersburg, 1885), 674-78.

66. Svechin, *Russko-yaponskaya voyna, 1904-1905 gg.*, 387.

67. Svechin, "Strategicheskiy ocherk," *Voyennyi sbornik*, no. 4 (April 1907), 68-69.

68. A. Svechin, "Otvetstvennost' i takticheskie zadachi," *Voyennaya nauka i revolyutsiya*, no. 2 (1921), 177-80.

69. In 1909 Neznamov had used a public lecture to identify the central changes in the art of military leadership, which were arising from the demands of mass, industrial war. Much of what Neznamov said was taken from German writings, especially Sigismund von Schlichting, but they were presented within a very Russian context. Neznamov redefined control (*upravlenie*) and initiative (*pochin*) so as to stress the role of the commander in imposing order from above in the form of his plan of action. A. Neznamov, *Sovremen-*

naya voyna: Deistviya polevoy armii, 2nd edition (Moscow, 1912), 13-17.

70. A. A. Svechin, *Strategiya*, 1st edition (Moscow, 1926), 9-10.

71. Ibid., 10.

72. Ibid., 32.

73. Ibid., 32-33.

74. Ibid., 52-53.

75. Ibid., 65.

76. Ibid., 66-114.

77. Ibid., 257-58.

78. Ibid., 24.

79. Ibid., 258-59.

80. Ibid., 268-69.

81. Ibid., 277.

82. Ibid., 278.

83. Ibid., 66.

84. Vladimir Melikov, *Marna—1914 goda. Visla—1920 goda. Smirna—1922 goda* (Moscow, 1928), 439-44.

85. *Akademiya im. M. V. Frunze* (Moscow, 1973), 98.

86. This situation is quite clear from contemporary publications, articles, and regulations. See I. Ivanov, "Voyenno-tekhnicheskaya literatura po voprosam kharaktera budushchey voyny i operativnogo iskusstva," *Voyna i revolyutsiya*, no. 2 (March-April 1934), 13-30; *Field Regulations of the Red Army* [1929] (Washington, DC, 1985); and USSR, Narodnyy Komissariat Oborony, *Vremennyy poleovoy ustav RKKA 1936* (PU 36) (Moscow, 1937).

87. *Akademiya General'nogo Shtaba: Istoriya Voyennoy ordenov Lenina i Suvroova I stepeni akademii General'nogo shtaba Vooruzhennykh Sil SSSR imeni K. E. Voroshilova*, 2nd edition, (Moscow, 1987), 22-24.

88. Frunze, *Izbrannye proizvedenniya*, II (Moscow, 1957), 35.

89. V. Triandafillov, "Materialy dlya zadachi na shtabnuyu voyennuyu igru," *Voyna i revolyutsiya*, no. 12 (December 1927), 31-45.

90. Korotkov, *Istoriya sovetskoy voyennoy mysli*, 121.

91. A. Svechin, *Klauzevits* (Moscow, 1934), 19. This introduction was done to make certain that no one missed the ideologically subversive tendencies in Svechin's writings.

92. A. Verkhovskiy, "Novaya i staraya shkola," no. 4 (April 1928), 100-101.

93. Ibid., 109.

94. N. Varfolomeev, "Strategiya v akademicheskoy postanovke," *Voyna i revolyutsiya*, no. 11 (November 1928), 83-84.

95. N. Varfolomeev, "Dvizheniye presleduiushchey armii k polyu reshitel'nogo srazheniya," *Revolyutsiya i voyna*, no. 13 (1921), 69-96; N. Varfolomeev, "Manevry na zapfronte," *Revolyutsiya i voyna*, no. 19 (1923), 5-26, and no. 20 (1923), 77-104; and N. Varfolomeev, "Strategicheskoye narastaniye i istoshcheniye v grazhdanskoy voyne," in A. S. Bubnov et al., eds., *Grazhdanskaya voyna 1918-1921: Voyennoye iskusstvo Krasnoy armii* (Moscow, 1928), 260-81.

96. Jacob W. Kipp, ed., *V. K. Triandafillov's "The Nature of the Operations of Contemporary Armies"* (London, 1994).

97. V. K. Triandafillov presents a "rightist" view in "Vozmozhnaya chislennost' budushchikh armiy," *Voyna i revolyutsiya*, no. 3 (1927), 14-43.

98. V. Triandafillov, *Kharakter operatsii sovremennykharmiy*, 3rd edition (Moscow, 1936), 7-9, 255. Triandafillov's study of the Perekop operation was later reworked and published as part of the three-volume history of the Civil War. See also N. Triandafillov, "Perekopskaya operatsiya Krasnoy armii (takticheskiy etyud)," in Bubnov et al., *Grazhdanskaya voyna 1918-1921: Boevaya zhizn' Krasnoy armii,* I, 339-57.

99. V. Triandafillov, "O Volkovysskoy operatisii," *Krasnaya Armiya: Vestnik Voyenno-Nauchnogo obshchestva pri Voyennoy Akademii,* nos. 10-11 (January-February 1922), 34-43.

100. V. Triandafillov, *Kharakter operatsiy sovremennykharmiy*, lst edition (Moscow: Gosizdat, Otdel Voyenlit, 1929), 1 ff.

101. *Field Regulations of the Red Army 1929* (Washington, DC, 1985), 63-93. The tie between future war (*budushchaya voyna*) and operational art (*operativnoe iskusstvo*) was made by I. Ivanov in a bibliography he published in 1934. There the posthumous second (1933) edition of Triandafillov's book was listed as the basic work in four out of twelve major categories, that is, contemporary operational means, the conduct of operations, meeting operations, and offensive operations. Under the subtopics listed for conduct of operations, *Kharakter operatsy sovremennykh armiy* was listed as the basic work for studying general questions, control of operations, and transport and rear. See I. Ivanov, "Voyenno-tekhnicheskaya literatura po voprosam kharaktera budushchey voyny i operativnogo iskusstva," *Voyna i revolyutsiya*, no. 2 (March-April 1934), 13-30.

102. B. M. Shaposhnikov, *Mozg armii, in Vospominaniya. Voyenno-nauchnye trudy* (Moscow, 1974), 425-29.

103. V. Triandafillov, *Kharakter operatsiy sovremennykharmii*, 3rd edition (Moscow, 1937), 235-54.

104. M. N. Tukhachevsky, "K voprosu o sovremennoy strategii," in *Voyna i voyennoye isskustvo v svete istoricheskogo materializma* (Moscow, 1927), 127-33.

105. G. Isserson, "Zapiski sovremennika o M. N. Tukhachevskom," *Voyenno-istoricheskiy zhurnal*, no. 4 (April 1964), 65-67.

106. USSR, RKKA, IV Upravlenie Shtaba, *Budushchaya voyna* (Moscow, 1928), i-vii.

107. Ibid, xi, xii.

108. Ibid., 724-35.

109. Ibid., 638.

110. Ibid., 645-46.

111. Ibid., 650.

112. Ibid., 653.

113. Ibid., 653-54.

114. Ibid., 656-57.

115. Ibid., 690.

116. Viktor Novitskiy, "Bor'ba za kharakter budushchey voyny," *Voyna i revolyutsiya*, no. 5 (May 1929), 1-13.

117. A. Lapchinskiy, "Deistviye aviatsii v nachal'nom periode voyny," *Voyna i revolyutsiya*, no. 6 (June 1929), 55-66; Ia. Alksnis,"Nachal'nyy period voiny," *Voyna i revolyutsiya*, no. 9 (September 1929), 3-22, and no. 10 (1929), 3-15; V. Novitskiy, "Deistviya aviatsii v nachal'nom periode voyny," *Voyna i revolyutsiya*, no. 9 (September 1929), 23-31; R. P. Eideman, "K voprosu o kharaktere nachal'nogo perioda voyny," *Voyna i revolyutsiya*, no. 8 (August 1931), 3-12; E. Shilovskiy, "Nachal'nyy period voyny," *Voyna i revolyutsiya*, nos. 9-10 (September-October 1933), 3-11; M. N. Tukhachevskiy,

"Kharakter pogranichnykh operatsiy," in *Izbrannye proizvedeniya*, II (Moscow, 1964), 212-21; S. N. Krasil'nikov, "Nachal'nyy period budushchey voiny," *Pravda* (May 20, 1936), 2; G. Isserson, *Novye formy bor'by* (Moscow, 1940); A. I. Starunin, "Operativnaya vnezapnost'," *Voyennaya mysl'*, no. 3 (March 1941), 27-35.

118. I. V. Stalin, "O zadachakh khozyyaystvennikov," in *Sochineniya*, XIII (Moscow, 1951), 39.

119. "Motorizatsiya i mekhanizatsiya inostrannykh armiy (k nachalu 1929 g.)," *Informatsionnyy sbornik*, no. 12 (December 1928), 145-57.

120. Tukhachevskiy, "Na baze dostignutogo—k novym zadacham," *Izbrannye proizvedeniya*, II, 67-68; and D. N. Nikishev, "Chelovek dela," in N. I. Koritsky et al., eds., *Marshal Tukhachevskiy: Vospominaniya druzey i soratnikov* (Moscow, 1965), 199-202.

121. *Sovetskaya voyennay a entsiklopediya*, II (Moscow, 1933), 842-43.

122. I. Stalin, "O rabotakh aprel'skogo oby'edinennogo plenuma TsK i TsKK," in *Sochineniya*, XI, 53-63.

123. Tukhachevskiy, "Predisloviye k knige G. Del'briuka *Istoriya voyennogo iskusstva v ramkakh politicheskoy istorii*," in *Izbrannye proizvedeniya*, II, 116-46.

124. For an extended and critical treatment of this military-political struggle, see N. I. Nikiforov, "Svechin—Tukhachevskiy," *Gepolitika i bezopasnost'*, no. 2 (1994), 72-80.

125. A. S. Bubnov, "Voennaya sektsiya i ee blizhaishiye zadachi," in Kommunisticheskaya Akademiya, Sektsiya po izucheniyu problem voyny, *Zapiski*, I, (1930), 5.

126. M. N. Tukhachevsky, "O kharaktere sovremennykh voyn v svete resheniy VI kongressa Kominterna," in *Zapiski* (Moscow, 1930), 21-29.

127. *Voyna i revolyutsiya*, no. 3 (1930), 140-47.

128. Stephen F. Cohen, "Bolshevism and Stalinism," in Robert C. Tucker, *Stalinism: Essays in Historical Interpretation* (New York, 1977), 3-29. See also James J. Schneider, *The Structure of Strategic Revolution: Total War and the Roots of the Soviet War State* (Novato, CA, 1994).

129. Tukhachevsky, "Predisloviye k knige Dzh. Fullera *Reformatsiya voyny*," *Izbrannye proizvedeniya*, II, 152.

130. Tukhachevky, "Novye voprosy voyny," in *Izbrannye proizvedenniya*, II, 184-87.

131. G. Isserson, *Evolyutsiya operativnogo iskusstva* (Moscow, 1932), 1 ff.

132. Ibid., 55-65.

5

Operational Art and the Canadian Army's Way of War

William McAndrew

This could be a commendably short chapter. Arguably, Canadian army commanders have never been in a position either to plan a campaign or to practice operational art. Count Frontenac was on the fringe in his time, but, possibly, the closest anyone came to independent command of a campaign was Major General Sir Fred Middleton in the Canadian northwest, if he can be appropriated as a temporary Canadian. During the wars of this century Canadians functioned only at the tactical level, under British commanders who may or may not have exercised operational art.

Canadian commanders in the Second World War, with which this chapter is concerned, would scarcely recognize the terminology that has evolved in recent decades, especially in publications from Leavenworth, Carlisle, and Camberley, that have introduced a fresh vocabulary to military studies.[1] As writers have defined that sphere of thought and action between the *strategic* and the *tactical*, a commander formulates a *concept of operations* to conduct an autonomous *campaign* in a *theater of war* and/or *theater of operations*. At least in modern times, it will likely be *joint*, that is, interservice, and probably *combined*, as part of a coalition.[2] The commander's concept will envisage whatever *sequential,* tactical engagements along *lines of operation* are needed to achieve his aim. His ultimate objective will be the enemy's *center of gravity*, usually identified as his main fighting force, but it could possibly be a decisive geographical area, or a vital command and communications center the loss of which would be paralyzing. The successful commander will ensure that he does not outrun his resources, or go beyond his *culminating point*, while trying to force his enemy beyond his. In less categorical times a commander described the process as getting to where he was going "fastest with the mostest."

A distinction should be made between campaign planning and operational art. The former is straightforward: a systematic, analytical process of getting from here to there, along the lines of an engineer's critical path to build a bridge. Operational art is less easily described: more a way of intuitive thinking, a facility to discern patterns in diversity, a continuing process rather than a finite end. Some writers assume implicitly that operational art conceptions have been discovered, as they float universally through human experience. To others, they are more prosaic intellectual inventions to order thought.[3]

Clausewitz is at the center of most recent studies, and most agree with Richard Swain that "the history of operational art is to be found in the accounts of campaigns and the independent actions of large units within a theater of operation or theater of war."[4] Within that framework, the primary focus of study, Jay Luvaas has written, must be men not maxims, commanders not the mechanics of war. He cites Napoleon that battle "has no precise, fixed rules. Everything depends on the character that nature has given to the general, on his qualities, on his faults, on the nature of the troops, on the range of weapons, on the season and on a thousand circumstances which are never the same."[5] Besides Napoleon, examples of operational art have been detected in Alexander's maneuvers, Ghengis Khan's sweeps, and Marlborough's marches, as well as in Clausewitz's mind.

If operational art is a way of thinking about war in universal terms—conceptions plucked out of the ether—it is not inconceivable that its insights lurked in the minds of Canadian generals—Currie, McNaughton, Crerar, Simonds, Burns, to take the most senior. If so, however, they do not seem to have recorded them, certainly not in terms now in common usage. If, in contrast, operational art is more a pragmatic invention, there is little reason to expect that Canadians would have embraced abstract operational ideas that were, in the modern era at least, essentially European, continental in scope, and conducted by massive conscript armies. Not only was there no practical scenario in which Canadians might have applied the precepts of operational art, they lacked any practicable means of implementing them.[6]

The interwar era was, ironically, a heyday of Canadian army planning. The successive Defence Schemes and papers prepared by Colonels Henry Crerar and Maurice Pope are closely reasoned models of rational exposition. But they were concerned with the mobilization of expeditionary forces that could readily be absorbed into the British army, not with operational art. Considering the constraints controlling their imaginations, this is unsurprising. The absence of clear strategic direction precluded a national military campaign plan that might have been informed by operational art. The unilateral defense of Canada's share of North America was an impracticable proposition, even if a large conscript army had at any time been available. Conceivably, operational insights would have been helpful to direct the fifteen-division army envisaged in Defence Scheme No. 1 for defense against the United States. Colonel Sutherland-Brown did envisage a maneuver campaign, to buy time until the combined and possibly joint might of Britain, Australia, and India arrived to intervene, but it may be doubted that Spokane, Fargo, Minneapolis, Albany, and Bangor—on which the Canadian Militia's "guerrilla swarms" were to converge—constituted decisive centers of gravity, the seizure of which would have crippled the United States.[7]

Canadian commanders thought realistically, within the limits of the possible, and the foremost reality was that their services were subsumed almost entirely within British organizations, technology, and doctrine. At the Staff and Imperial Defence Colleges Canadians absorbed both the objectives and styles of British planning that, themselves, were hamstrung at the strategic level by the lack of an unequivocal focus. Imperial interests were not easily harmonized with continental commitments.[8] Canadian independent planning in the interwar years was comparably skewed between national defense needs and military predilections for participating in an expeditionary force overseas. In that milieu, Canadian military commanders had little practical need to think about operational art, even if they had been so inclined. Steve Harris has described the result:

> Had the army been given a clear commitment to a role in which imperial cooperation was unimportant, it might have produced more independent military thought in Canada, as officers would have been forced to prepare for operations they would both plan and direct. But that called for an uncharacteristically firm and focused decision on defence policy by the government. Moreover, most of these non-imperial contingencies were so hypothetical that they did not lend themselves to serious consideration and rigorous study: by 1930 the United States had long since been abandoned as a potential enemy; there was minimal political support for military service in aid of the League of Nations; and invasion from overseas was unthinkable. Left with just one *raison d'etre*—to fight alongside the British army in a major war—the general staff drew the obvious conclusion. Operational planning would be in the hands of the British who would control the Canadian army once war broke out, and British doctrine was accepted as given.[9]

Echoes haunt current realities. One important effect of the consequent abandonment of operational direction was the loss of any possibility of national joint action when war broke out in 1939. Political and military dynamics sent the navy, army, and air force off in separate directions, with each service left to function as quite different but subordinate coalition partners. Considering that by the end of the war, from a population of scarcely twelve million, Canada had about one million men and women in uniform, manning a field army of five divisions, almost 100 air force squadrons, and 400 naval vessels, the absence of coherent national command and control can be viewed as either absurd or scandalous.[10]

Besides practical obstacles, there is little indication that Canadian commanders aspired to a level of abstraction in intellectual fields that might have led them to consider the nuances of operational art. This was a crippling deficiency if it is essentially a way of thinking about and conceptualizing military affairs.[11] Practical men in a peacetime era managing restricted budgets, low troop levels, and obsolescent equipment, the commanders had no opportunity to exercise large formations directly. This meant, then as now, that they could discover operational art only through historical study of campaigns and commanders. Unfortunately, then as now, historical-mindedness was an uncommon currency, and historical precedents did not extend much beyond shared personal experiences of the Western Front. If Canadian commanders thought in

Clausewitzian patterns they did not parade it. A recent study remarks about Clausewitz that "His approach has more in common with that of the well-schooled art critic than with that of the physical scientist, for where the scientist seeks to set forth propositions verifiable by experiments that can be duplicated, the critic seeks to understand unique events."[12] Inclined to a straightforward, analytical approach to military affairs, Canadian commanders did not gravitate naturally to the intuitive insight which informs operational art.[13]

While Canadian commanders lacked the need, means, and inclination to think or function at the operational level in the Second World War, they certainly felt the effects of others' use and misuse of its concepts. This is particularly evident at the junction at the top with strategy, from which Canadians were excluded, and at the bottom with tactics, with which they were very much concerned. That joint between the operational and tactical levels would seem to be a vital factor in military effectiveness. Mere tactics themselves may not win wars, but the purest operational conception will remain barren if the tactical means to implement it are deficient. Several factors link the tactical and operational levels, for example, troop quality, organizations, and technology. The most vital ingredient, however, would seem to be doctrine; not doctrine as dogma, but simply the shared premises, assumptions, and procedures that allowed soldiers, units, and formations to function as a coherent whole.

If doctrine is the chain linking the tactical and operational levels, its essence is worth exploring; and that is most conveniently done, in familiar fashion, by comparing German and British/Canadian practices. As has been amply described in recent years, First World War experiences cast different doctrinal shadows on the losers and winners. Each drew on contrary lessons that shaped their responses when they resumed fighting in 1939. The German application of *auftragstaktik* had loosened the tactical battlefield of the trenches in 1918, and they adapted it in the interwar years into blitzkrieg, which could exploit an open battlefield. That mission-directed way of war, emphasizing delegation of responsibility and the maximum exercise of initiative, harmonized tactics with the operational level. Reinforced by an appropriate reward system that acknowledged independent decisionmaking,[14] *auftragstaktik* required constant training in a way of thinking to instill and sustain the mutual trust underlying it.[15] "From my first day as a student officer," one well-practiced panzer-grenadier officer has written,

> the expression "Repeat the Mission" rang in my ears. Our superiors wanted us to "repeat the mission" that had been assigned us to be quite sure that we understood it. And they always said "Mission" and not "Order". . . . And I always did the same with my subordinates to whom I always passed on the "Mission". . . . with *auftragstaktik* a mission is ordered and the officer is left with the freedom to carry out the mission assigned to him, and so he feels responsible for the actions which are suggested to him by his intelligence, his enterprise and his capabilities.[16]

Dynamic doctrine allowed the lowest-level tactical unit to share in the thought process of the highest-level operational commander. While Germans acknowledged an inherently chaotic battlefield, and used *auftragstatik* as a means to exploit it, Canadian and British experience of the First World War led them in other directions. Rather than accepting the essentially unpredictable nature of the battlefield, they tried to impose order on it. Artillery had dominated their war and their expertise lay in refining its effective management. Quite naturally, firepower and technology prevailed over finesse. Instead of being fluid, Canadian/British doctrine was essentially static.[17] Rather than directing formations along paths of least resistance to deep operational centers, doctrine perfected the set-piece, or deliberate, attack for tactical goals.

Control produced a rigid, centralized management style. Higher-level staffs carefully crafted detailed plans for others to implement. Divisions, brigades, and battalions were routinely assigned limited tactical objectives, invariably a geographical feature which was usually an enemy strong point. Start lines, report lines, boundaries, and timed artillery barrages gridded the battlefield, confining tactical mobility, let alone operational maneuver, and leaving unit commanders little opportunity to respond flexibly. Two examples of this way of war, in Italy and in northwest Europe, may illustrate the pattern.

Besides its contentious strategic intent, about which the coalition leadership strongly disagreed, the Italian campaign offers a classic study of the disjunction between the operational and tactical levels of war. Operational intentions echoed Vivaldi in their seasonal variations. The initial aims of freeing the Mediterranean for shipping, and knocking Italy from the war, gave way to one of securing air bases in the Foggia plains from which strategic bombers could strike Central Europe. Finally, the campaign evolved into a gigantic holding operation, whose purpose was to keep German divisions away from Normandy and the Eastern Front. Unwittingly or not, this operational aim became progressively detached from the many subsidiary tactical objectives up the Italian boot until the two worked at cross-purposes. While the eventual operational purpose was to keep German divisions in Italy, the tactical aim was to drive them back to the Alps and beyond. As Sir David Hunt has written, "The paradox of the Italian campaign lies in the fact that [tactical] failure was the means by which the Allied Armies in Italy succeeded in fulfilling their strategic purpose."[18] It is no coincidence that the Italian campaign conceived *Catch-22*.

Possibilities for operational success eluded the Allies at several stages. Coalition command differences and an appalling absence of joint service intervention permitted the Germans to withdraw from Sicily across the Strait of Messina. Lack of resources, and possibly of imagination, severely limited Allied amphibious action, which the Germans feared most.[19] The result was an interminable tactical slog against the topographical grain. Sir Michael Howard has remarked, "The lesson that most of those involved in the campaign on either side would agree about, is that if one wants to conquer Italy, the southern end of the peninsula is not the best place to begin."[20] Canadians who fought from Pachino to Ravenna likely agreed, as they felt the consequences of disjunction.

Their experience also raises questions about the compatibility of their way of war with operational art; a good example of these problems is provided by the Battle for Rome in the early summer of 1944.[21]

The operational aim of the Allied offensive toward Rome was to trap and destroy the German Tenth Army, in two decisive tactical encounters. Following an intensive air program to interdict German supply routes,[22] the British Eighth Army, along with the Second U.S. and the French Expeditionary Corps, were to attack northwest from the Gustav Line below Cassino. After luring German reserves to stop them, the Sixth U.S. Corps was to break out from its Anzio beachhead and meet the southern pincer.

The main task of General Burn's First Canadian Corps in the battle was to clear the ground between the Gustav and Hitler Lines, break through the latter, then exploit along the Liri valley toward Rome. When First Division's leading units reached the Hitler Line they stopped, as was standard doctrinal practice, to regroup, move guns forward, pull in the logistics tail, and plan a set-piece attack. Relying on massive firepower, Eighth Army's attack devolved upon three of First Division's battalions. Using standard two-up and one-back formations at all levels, this spread pointed riflemen at ten-meter intervals across the two-kilometer front. They were to follow an 800-gun artillery barrage that would roll methodically in front of them on a timed program. When they had carved a gap, Fifth Armoured Division was to go through to exploit.

Stacking ammunition, preparing counterbattery programs, and arranging the exquisite details of the barrage took three days, during which time the Germans strengthened their defenses. In the meantime, North African colonial troops in the French Expeditionary Corps (FEC) had marched over what were thought to be impassable mountains in an unorthodox, yet classic, operational maneuver, and outflanked the Hitler Line on the Canadians' left. Their commander, Marshal Juin, tried to persuade General Alexander to allow him to keep going to trap the Germans, but the interarmy boundary was in the way and Alexander demurred. Meant to control tactical movement, the boundary had the unfortunate effect of hampering operational maneuver.

Although the need for their attack was now problematical, Second Brigade with the Carleton and York attached was ordered nonetheless to assault at first light on 23 May. The evening before, the commanding officers of two of Second Brigade's battalions became bothered that the plan they had been given to implement was too rigid and left them no flexibility. They were particularly concerned that their leading companies would be unable to conform to the timings of the barrage in the rough terrain. One unit history recounts that "they decided to consult their Brigadier; at his Headquarters they found he had retired. He did not emerge from his dug-out but a sleepy voice informed the officers that no further changes could be made."[23]

Selection and maintenance of the plan, not the aim, being the guiding principle, the two COs made what coordinating arrangements they could between themselves before duly sending their men into an inferno a few hours later. Their worst apprehensions were realized when the barrage moved on according to its fixed schedule; supporting tanks were stopped by mines, and riflemen were left to inch their way through wire. Even worse, German paratroopers, on abandoning Cassino, moved into position alongside a wide ravine on Second Brigade's right flank that was just beyond the barrage's rectangular perimeter. From there they took the riflemen under devastating enfilade fire that inflicted many of the brigade's 540 casualties.

The battle was fought according to doctrine.[24] The meticulous fire plan delivered to batteries the evening before the assault was a technical masterpiece, accounting for all foreseeable contingencies except the most important one of being flexibly coordinated with the movement of the infantry they were meant to support. Infantry movement, from report line to report line, was restricted to timings that discounted the inherent friction and confusion of the battlefield, any battlefield. Not knowing that the Princes Patricia's Canadian Light Infantry (PPCLI) in front of him had been stopped, when communications broke down—as they invariably did—the commanding officer of the Loyal Edmontons, in reserve, felt obliged to commit his battalion according to the prearranged timings controlling the overall plan. His companies simply ran up the backs of the Patricias, to be slaughtered like them.[25]

Fortunately, the left assault battalion, the Carleton and York, found a weak spot in the defenses, and next day Fifth Armoured went through the gap. Unfortunately, after a quick start, it bogged down, along with the rest of Eighth Army, and the offensive's operational objective began to fade. This was not unusual. The unsuitability of doctrine to operational maneuver was noted in a 1944 British pamphlet. "Our own tactical methods are thorough and methodical but slow and cumbersome," it noted.[26] This proved the case for the Canadians. With two corps scrambling for scant track space on a narrow front, traffic congestion by 20,000 vehicles provided a vivid illustration that wheels did not mean mobility.[27] Nor could mobility be transposed into operational maneuver. As one recent observer notes, while maneuver is not essential to operational art, it depends heavily on it, so long as "manoeuvre is a frame of mind . . . more than [mere] mobility or movement. Mobility is the means by which we execute maneuver. Only when mobility is applied with the aim of inflicting paralysis on the enemy is it translated into manoeuvre. Manoeuvre is not a way of moving but a way of thinking."[28] In this case, the force that could maneuver operationally, the FEC, was kept on a short lease, while Eighth Army pressed methodically toward Rome at an average speed of about five kilometers a day. Then, when General Mark Clark decided to send his Anzio force directly on Rome, rather than complete his assigned move, the operational objective faded completely as the Germans withdrew to fight another day, and week, and on for months.

In contrast to the Allied effort Field Marshal Kesselring's mission for the withdrawal is a prime example of clarity and simplicity in operational direction:

> Withdraw fighting, bring into the line of battle from the rear and from the flanks the reserves already on the march southwards, close gaps between the various units, and build up the internal flanks of the units themselves . . . after the major formations in crisis have been re-ordered, halt and concentrate on defensive positions, as far south as possible.[29]

The Allies obliged this effort.

Comparable operational lapses characterized the Allied campaign in northwest Europe. Several factors intervened. Imprecise strategic direction dispersed the campaign's operational effort, and General Eisenhower's command structure was seriously distorted. Jointness was fatally flawed, with the army and air force differing over roles, and failing to apply their combined bomber and fighter resources effectively where and when they were most needed.[30] Then, following the eventually successful attritional struggle in Normandy, commanders were unable to agree about the enemy's center of gravity—German morale, the German army, the Ruhr, the Saar, Berlin—and about whether to reach whichever on a broad or narrow front. In the event itself, the most feasible operational objectives—the junction points of the disorganized withdrawing German armies near Trier, Aachen, and Antwerp—were lost.

As in Italy, Canadians in northwest Europe functioned only on the tactical level, but they were at the tantalizing operational fringe at Antwerp. The story is by now shopworn. British troops reached Brussels on 2 September and next day drove onto Antwerp's intact docks. Despite the urging of a gallant Belgian underground leader, the leading troops declined to seize a nearby bridge over the Albert Canal and the Germans recovered. The immediate commanders involved later agonized over their failure to seize canal crossings, seal off the Beveland Peninsula, and drive north when few Germans blocked the way. Antwerp was, arguably, an operational objective, both because of its port's vital logistical significance, and because the sector was the seam between the withdrawing German Fifteenth and the still-organizing First Parachute Armies. Kept apart, each might well have been defeated in detail, but commanders at all levels let the incomparable opportunity pass. The Fifteenth Army withdrew from the Scheldt and Walcheren, and strengthened a new defense line; the Canadians were doomed to miserable and costly tactical operations clearing Antwerp's approaches.

Given their common doctrinal limitations, there is little reason to expect that Canadian commanders would have reacted any differently from their British counterparts. The Canadian task in the pursuit was to secure the ports strung along the English Channel by medieval-like siege tactics. Well after the war, General Simonds claimed that he had proposed bypassing the ports and sweeping boldly to Antwerp, but this is difficult to confirm from contemporary documents.[31] In any case his Second Corps advanced methodically, from bound to bound, and it is unlikely that doctrinal style would have tolerated, let alone encouraged, bold sweeps. One reconnaissance unit commander complained during the move that he was being slowed unnecessarily by his orders, that he was continually running beyond his assigned limited objectives. Instead of fighting units keeping close behind him as he cleared routes, commanders ordered his scout cars to withdraw each evening to laagers 20-30 kilometers in the rear.[32] At a divisional conference, a brigade commander remarked that he had "too little room for manoeuvre, the flanks being too close to the axis." Another observed that his units were not mobile enough to outflank the withdrawing enemy, who he said, "was merely being driven ahead." The divisional commander's response accurately reflects the doctrinal constraints. In his view ". . . it was not essential that we destroy the enemy in the course of our pursuit. . . . [O]ur function [is] little more than that of 'beaters' flushing the enemy out of cover and into the final killing-ground."[33]

There was no killing ground, and this was hardly a way of thinking to implement tactically a higher commander's operational intent, assuming that he had one. The Germans got away to fight through the winter, and disunity lost an incomparable opportunity in the vital first week of September. With Americans on the German border, and British troops heading east from Antwerp, centers of gravity dispersed further. The Ruhr remained vaguely in the background, although how long it would take for its isolation to paralyze Germany is not clear. Neutralizing V-bomb sites in Holland became a prime tactical objective, as did gaining a bridgehead across the Rhine. Instead of combining resources in a concentrated thrust east Montgomery went north to Arhnem.[34]

Failure in autumn led to First Canadian Army's largest engagement of the war, the Battle for the Rhineland in the early spring of 1945. The tactical battle itself was made necessary by a failure to take the operational objective of the Roer River dams in September. At that time the dams on the Roer River that controlled the flow of water into the Rhineland were virtually undefended, but commanders unfortunately failed to appreciate their significance, leaving a stern legacy for the following spring. The Allied offensive was initially planned as a two-pronged assault; the Ninth U.S. Army to move from the south, and First Canadian Army from the north, to trap the defenders between them. However, by flooding the Roer valley with a controlled flow of water from the dams, the Germans were able to dictate the pace of Rhineland operations. Instead of having to split their defenses against the coordinated pincers, they were able to concentrate their forces against the Canadian attack, Operation Veritable, when it began in early February.

Operation Veritable was the epitome of the Canadian army's way of war. In an area of about 80 by 40 kilometers, General Crerar deployed resources for his set-piece attack that were greater than those at Vimy Ridge, and not far behind those for Overlord. With ample time to prepare, staffs carefully arranged the battlefield in impeccable detail. The sheer scale and scope of their planning is worth noting, if for no other reason than a Canadian army is unlikely ever to see it again. General Crerar's two corps—Second Canadian and Thirtieth British —included seven infantry and three armored divisions, with three independent armored brigades. Besides organic divisional field regiments, there were one super heavy, five heavy, and 17 medium regiments, as well as the Canadian Rocket Battery, totalling about 1,200 guns. Around 350 different ammunition types were required, the portion shot between D-3 and D-Day being equivalent to 25,000 medium bomb loads. There were 3,400 tanks, with a quarter-million rounds for their main armament, along with Crocodiles, Buffaloes, Weasels, Kangaroos, and Wasps, all spread behind a 50-kilometer continuous smoke screen. The army's ration strength was more than 470,000, most of them not Canadian. Sappers constructed and widened 170 kilometers of roads, using 18,000 logs, on which 35,000 vehicles drove an average of 200 kilometers daily, using five million liters of gasoline in the buildup phase. Sappers also had 2,000 tons of bridging equipment, including a Bailey span of 400 meters. There were 446 freight trains using 13 railheads which dumped tonnage equal to 89,000 three-ton vehicle loads. One thousand soldiers deployed on traffic control duties planted 9,600 route signs. More than 500,000 air photos, 15,000 enlargements, and 819,000 map sheets used 31 tons of paper. Air support was equally impressive. General Crerar could call on the resources of both 84 and 83 Groups of Second Tactical Air Force, the latter for air cover, the former for close support, as well as over 1,000 heavy and 200 medium bombers.[35]

Planning the battle was a prodigious achievement that displayed Canadian military staff bureaucracy at its best, highlighting all the strengths of conventional doctrine for the set-piece attack: meticulous staff work, massive firepower, careful movement. Once under way, however, the battle inevitably demonstrated the truism that plans seldom survive first contact with the enemy. That evening the Germans breached more Rhine dikes submerging much of the main supply route. Thousands of vehicles turned thawing ground into impenetrable mud, and, when reserve troops were committed prematurely, movement stopped. One of the participants described

> a day of nightmare traffic congestion which made coherent troop movement almost impossible. In fact it had been proved—it is to be hoped for all time — that two divisions cannot operate satisfactorily on one axis—especially when the axis itself leads through a bog and is itself in places under water. In the circumstances it is not surprising that throughout the day the contacts of many of the commanders involved had been of a character which cannot justly be described as being noteworthy for their cordiality.[36]

The Germans gained time to reinforce, and leading troops were stopped in Cleve by rubble caused by the bombing that had been meant to ease their way. Ground, weather, Germans, and doctrine combined to prevent General Crerar from applying the abundant resources he had available at any one vital point. The usual attritional struggle followed.

These isolated, but typical, tactical encounters reveal a pattern of thinking, a style, a way of war, with an inherent fault at the junction between the operational and tactical levels. In theory, Canadian doctrine encouraged initiative and flexibility; however, if unit and personal accounts are at all indicative, actual practice was the opposite. Instead of a common, unbroken thread connecting all levels of command, several staff layers intervened between planners and implementers. Rather than being delegated, responsibility was centralized, and the execution of operations was made highly dependent on rigid orders and detailed plans delivered from above.[37] Bureaucratic order and managerial competence prevailed over creative imagination. Instead of synchronizing tactics with operational insight, doctrine—the way of war—got in the way.

There is little reason to suppose that the pattern of thinking has changed in the intervening years. Emphasis on management, staff bureaucracy, and top-down direction mirrors Canada's other institutional structures: social, economic, cultural. Without a profound institutional shock, an army is unlikely to change its style. The U.S. Army was shocked out of its Second World War rut by Vietnam and, while searching for its collective soul, rediscovered the operational level of war. That stimulating search through historical precedents now seems to have entered a new stage, in which insights congeal into antithetical conventional wisdom. Axioms codify into principles, and principles to checklists, which surreptitiously replace thought. If so, it seems likely that the urge to conformity will defeat the intent that gave the original quest its legitimacy. It is anything but clear that the vitality of this provocative intellectual revolution will be continually regenerated, or if it will degenerate, as usual, into bureaucratic impasse.

The Canadian Forces have not experienced that vital intellectual search for first principles. Instead of stimulating an exchange of ideas on which to construct a sound intellectual base, a bureaucracy arbitrarily directed that operational art was to be adopted. Unfortunately, this came at the time when, elsewhere, categories were hardening and insights were being engraved in doctrinal manuals. Accepting those manuals without having experienced, or really understood, the essential first phase builds on a precarious foundation. "Armies, like nations," Richard Hooker has remarked, "have cultures which profoundly influence their behavior. To change the way armies fight, one must begin not with field manuals, but with the way an army thinks about itself."[38] It is doubtful that a way of thinking can be changed by fiat, nor is it likely that the way an army thinks about itself can be imported. Trying to absorb foreign doctrines secondhand will be as fruitless as transplanting tropical plants in the tundra.

Assuming there is a need to preserve, or formulate, a corporate war-fighting memory, a profound educational challenge faces all levels of military instruction, from corporal to general. Staff colleges can readily teach the science and mechanics of campaign planning, but to be effective it will have to be solidly linked with tactical thinking by suitable doctrine. Moreover, if operational art says anything to higher commanders, its practice can only come from serious study of historical experience. There are no battle-wise commanders, and it is difficult to see how that professional insight can be gained other than vicariously through the lives of others. The American example can be eminently useful in this regard by demonstrating how the highest-level commanders can not only tolerate, but participate in, continuing historical learning, while putting to use the creative intellectual energy that is readily available in uniform. Commanders without actual experience can participate vicariously in military operations only through historical study. They may do this implicitly—their training and doctrinal manuals, even computer simulations, are based on someone's historical experience—or preferably through conscious study.[39]

As currently taught, war-fighting scenarios echo the 1930s. Lacking clear strategic direction, like then, operational control of Canadians now will also lie elsewhere, in combined forces directed against as yet undesignated enemies. This may be realistic, but the implications should be clear. Unless the Canadian Forces abandon any pretense of national sovereignty and distinct military autonomy, and are content simply to be absorbed as a few brigades, squadrons, and vessels into grand coalitions, a serious search for first principles is overdue. Otherwise there can be little apparent justification for an expensive military establishment.

The relationship of operational art to other-than-war situations also demands study. Clausewitzian insights can surely frame responses to quasi-war events, but just how they may best inform them is less clear.[40] Capriciously applying fashionable operational level of war precepts, as they are now being promulgated, to peacekeeping and peacemaking scenarios situates this appreciation. They may well be relevant, but in what particular ways is not necessarily self-evident. Other useful conceptual models may be out there waiting to be discovered or invented. Their essentials will only be revealed through sustained thought and open discussion.

NOTES

1. Lively dialogues have filled the volumes of *Military Review* and *Parameters*, from Leavenworth and Carlisle, over the past two decades. More recently, the British army introduced a three-month Higher Command and Staff Course to study the operational level. Several volumes of course papers have been published, for example, see Major General J. J. G. Mackenzie and Brian Holden Reid, eds., *The British Army and the Operational Level of War* (London, 1989). See also Edward N. Luttwak, "The Operational Level of War," in Luttwak, *Strategy and History: Collected Essays, Vol.2* (New Brunswick, NJ, 1980), 175-94.

2. For interesting naval views on the applicability of what originally were army conceptions to sea warfare see pseud."Trog Trog," "The Operational Level of Command in the Royal Navy," *Naval Review* 81 (1993), 211-20; and a rejoinder by Commander Rhys-Jones, "Correspondence," ibid., 419.

3. The two most succinct statements of operational art and campaign planning are in two U.S. Marine Corps publications: *Warfighting* (FMFM 1), and *Campaigning* (FMFM 1-1). For a more recent critique of these statements, see Major Philip E. Knobel, "Revise FMFM 1, Warfighting," *Marine Corps Gazette* (October 1993), 31-33; and Major Robert

S. Trout, "Dysfunctional Doctrine: The Marine Corps and FMFM 1, Warfighting," ibid., 33-35.

4. Colonel R. M. Swain, "The Written History of Operational Art," *Military Review* 70 (1990), 100-105.

5. Jay Luvaas, "Thinking at the Operational Level," *Parameters* 16 (1986), 2-6.

6. A critique of the interwar era is in John A. English, *The Canadian Army and the Normandy Campaign* (New York, 1991). Biographical background is in J. L. Granatstein, *The Generals: The Canadian Army's Senior Commanders in the Second World War* (Toronto, 1993).

7. Excerpts from Defence Scheme No. 1 are in James Eayrs, *In Defence of Canada: From the Great War to the Great Depression* (Toronto, 1964), 323-28.

8. As one observer has remarked, "[T]he problem of the British Army in the 1930s was not so much that British governments failed to build forces appropriate to their view of the world, than that their view was widely askew." David Schoenbaum, "Correspondence: Clausewitz and the British Generals," *International Security* 6 (Winter 1981-1982), 226.

9. Stephen J. Harris, *Canadian Brass: The Making of a Professional Army, 1860-1939* (Toronto, 1988), 205. Harris comments further "that the general staff's keen interest in protecting its mobilization plans had actually weakened the army's potential as a fighting force. Having concentrated on departmental organization and contingency planning as the means of achieving professional independence, the staff had little time or inclination to worry about how the army should be prepared for war. It forgot that professional soldiers were not just civilians in uniform." Ibid., 191.

10. The most comprehensive discussion of Canadian command policies during the war is in C. P. Stacey, *Arms, Men and Governments: The War Policies of Canada, 1939-1945* (Ottawa, 1970).

11. Canadian and British preference was for principles, codified like the commandments into ten immutable ones. See J. I. Alger, *The Quest for Victory: The History of the Principles of War* (Westport, CT, 1982); and Zvi Lanir, "The Principles of War and Military Thinking," *The Journal of Strategic Studies* 16 (1993), 1-17.

12. Eliot A. Cohen and John Gooch, *Military Misfortunes: The Anatomy of Failure in War* (New York, 1990), 44-45.

13. Nor did American commanders, according to some historians. See Martin Blumenson, "A Deaf Ear to Clausewitz: Allied Operational Objectives in World War II," *Parameters* 23 (1993), 16-27.

14. In *Fighting Power* (Westport, CT, 1982), 109-10, Martin van Creveld describes an incident illustrating how the Germans awarded decorations to commanders, not for physical bravery, but for exercising independent command judgement.

15. On First World War developments, see Bruce I. Gudmundsson, *Stormtroop Tactics: Innovation in the German Army, 1914-1918* (New York, 1989). For a German analysis of American doctrine and practice, see General Franz Halder, et al., *Analysis of U.S. Army Field Service Regulations*, Historical Division, United States Army, Europe, 1953.

16. Gerhard Muhm, "German Tactics in the Italian Campaign," copy in author's possession. See also W. Heinemann, "The Development of German Armoured Forces 1918-40," in J. P. Harris and F. H. Toase, *Armoured Warfare* (London, 1990), 51-69; and the papers edited by the Militaergeschichtliches Forschungsamt Freiburg im Breisgau in *Development, Planning and Realization of Operational Conceptions in World Wars I and*

II (Herford and Bonn, 1989). For a critique, see Roger Beaumont, "Wehrmacht Mystique Revisited," *Military Review* 70 (1990).

17. Differences between proponents of attrition and maneuver are part of a continuing debate. See, for example, John Ellis, *Brute Force: Allied Strategy and Tactics in the Second World War* (London, 1990); G. D. Sheffield, "Blitzkrieg and Attrition: Land Operations in Europe, 1914-45," in Colin McInnes and G. D. Sheffield, *Warfare in the Twentieth Century: Theory and Practice* (London, 1988), 51-79; "Commentary and Reply," *Parameters* 23 (1993), 107-10; Dominick Graham, "Observations on the Dialectics of British Tactics, 1904-45," in Ronald Haycock and Keith Neilson, *Men, Machines, and War* (Waterloo, Ontario, 1988), 49-73; Major George A. Higgins, "German and US Operational Art: A Contrast in Manoeuvre," *Military Review* 65 (1985), 22-29; and William J. McAndrew, "Fire or Movement? Canadian Tactical Doctrine, Sicily—1943," *Military Affairs* 51 (1987), 140-45.

18. Sir David Hunt, *A Don at War* (London, 1966). See also Hunt, "British Military Planning and Aims in 1944," in William Deakin, Elisabeth Barker, and Jonathan Chadwick, eds., *British Political and Military Strategy in Central, Eastern and Southern Europe in 1944* (New York, 1988), 1-20; and Brian Holden Reid, "The Italian Campaign, 1943-45: A Reappraisal of Allied Generalship," *The Journal of Strategic Studies* 13 (1990), 128-61.

19. The amphibious assault at Anzio in January 1944 was an exception, but its initial promise faded quickly.

20. Sir Michael Howard, *The Lessons of History* (New Haven, CT, 1991), 10.

21. General accounts include John Ellis, *Cassino: The Hollow Victory* (New York, 1984); and Raleigh Trevelyan, *Rome '44: The Battle for the Eternal City* (New York, 1982). The Canadian official history is G. W. L. Nicholson, *The Canadians in Italy, 1943-1945*, (Ottawa, 1966).

22. Air commanders differed over whether they should concentrate their attacks on main rail centers, all rail centers, or roads and railway tracks. The result of the lack of focus was that the interdiction program, Operation Strangle, was only partially successful in blocking German movements of supplies and reinforcements. See F. M. Sallagar, "Operation Strangle, (Italy, Spring 1944): A Case Study of Tactical Air Interdiction," Rand Corporation, 1972.

23. G. R. Stevens, *A City Goes to War* (Edmonton, 1964), 293.

24. On occasion breaking the doctrinal mould gained unexpected success, as a low-level tactical instance illustrates. On its approach to the Hitler Line the Forty-eighth Highlanders, moving in open formation astride a road from one report line to the next, came under fire from the right flank. Although not under fire himself, the lead company commander on the left stopped, according to doctrine, until the line could be straightened out and the situation clarified. Fortunately for the battle, if not for doctrine, the leading platoon was out of communication, so it kept going, got behind the withdrawing enemy, and secured an intact bridge that considerably speeded up the advance. At first highly annoyed at his temporary loss of control that the breach of doctrine seemed to produce, the commanding officer later acknowledged unorthodoxy by ensuring that the platoon commander and his sergeant were decorated. Kim Beattie, *Dileas: History of the 48th Highlanders of Canada, 1929-1956* (Toronto, 1957), 522-38.

25. The brigade commander faulted his battalion commanders for being too far forward, thus leaving him out of touch, but radio logs indicate that communications were maintained throughout. With another doctrine, of course, the brigadier would have been

sufficiently forward himself to influence the battle. Documentation is in National Archives of Canada (hereafter NAC), Ottawa, Ontario, RG 24, volumes 10,788, 10,881, and 10,922.

26. "In consequence our troops fight well in defence and our set-piece attacks are usually successful, but it is not unfair to say that through lack of enterprise in exploitation we seldom reap the full benefits of them." Directorate of History (hereafter DHist), Ottawa, Ontario, "Notes from Theatre of War, No. 20, Italy, 1943-44."

27. A daily average of 8,000 vehicles passed a checkpoint on the only main road, highway No. 6; on one day there were 11,542. See the official British history, C. J. C. Molony et al., *The Mediterranean and the Middle East*, VI, Part 1 (London, 1984).

28. A. Marcy, "Operational Art: Getting Started," *Military Review* 70 (1990), 107.

29. Quoted in Gerhard Muhm, "German Tactics in the Italian Campaign." See also William J. McAndrew, "Eighth Army at the Gothic Line: Commanders and Plans," and "Eighth Army at the Gothic Line: The Dogfight," *RUSI Journal* 131 (1986), 50-62.

30. This applies both to the heavy bomber forces and their dispersion of effort between oil, industrial, and city population targets, and the fighter dispersion between combat air patrols, armed reconnaissance, and ground support missions. For further discussion, see William J. (Bill) McAndrew, "Operational Art and the Northwest European Theatre of War, 1944," *Canadian Defence Quarterly* 21 (1991), 19-26.

31. In *Tug of War* (Toronto, 1984), W. Denis and Shelagh Whittaker cite postwar correspondence in making a case that Simonds proposed a wide sweep to Antwerp. In *The Long Left Flank*, (Toronto, 1988) Jeffery Williams discounts this as being improbable. Simonds's biographer, Dominick Graham, skirts the issue in *The Price of Command: A Biography of General Guy Simonds* (Toronto, 1993).

32. Summaries of unit actions are in DHist, 141.4A18013 (D6).

33. The divisional historical officer noted the discussions in his diary. See NAC, RG 24, volume 17,506.

34. See, for example, Richard Lamb, *Montgomery in Europe, 1943-45: Success or Failure* (London, 1983).

35. The most complete discussion is in John A. Macdonald, "In Search of Veritable: Training the Canadian Army Staff Officer, 1899 to 1945," M.A. thesis, Royal Military College of Canada, (Kingston, ON, 1992).

36. Major General H. Essame, *The 43rd Wessex Division at War, 1944-1945* (London, 1952), 206.

37. Such comments recur frequently in questionnaires about their battle experience completed by junior commanders. They are found in NAC, RG 24, volume 10,450. An earlier critic, who drew on his experiences in the Spanish Civil War, is Tom Wintringham. See *Deadlock War* (London, 1940). See also the discussion on fire and movement tactics, differentiating between infiltration and set-piece attacks, in "Minutes of 1st Battle School Conference Held at the Horse Guards on 17 and 18 Jun 42," NAC, RG 24, volume 9,764; and the "Record of Discussions of the Infantry Training Conference Held at the School of Infantry, Barnard Castle, 20-24 April 1944," in DHist 171.009 (D160). "The difference in conception and execution between these two tactics is fundamental. The first exalts the soldiers' intelligence and capability, the second tends to damp them down, making the soldier a passive executor of the orders of others." Gerhard Muhm, "German Tactics in the Italian Campaign."

38. Richard D. Hooker, "Attritionists—or Technologists?—vs Maneuverists," *Parameters* 23 (1993), 107-10.

39. One valuable forum of study is the battlefield tour, or staff ride. The author has conducted operational level study-tours of European battlefields with Canadian staffs at CENTAG/4ATAF over a period of ten years. Canadians could well keep in mind General Cushman's observation that "soldierly virtues such as integrity, courage, loyalty, and steadfastness are valuable indeed, but they are often not accompanied by insight. Insight comes from a willing openness to a variety of stimuli, from intellectual curiosity, from observations and reflection, from continuous evaluation and testing, from conversations and discussions, from review of assumptions, from listening to the views of outsiders, from a study of history, and from the indispensable ingredient of humility. Analysis, including systems analysis, can contribute to insight but it cannot substitute for it." John H. Cushman, "Challenge and Response at the Operational and Tactical levels, 1914-45," in Alan Millett and Williamson Murray, eds., *Military Effectiveness,* III (Boston, 1988), 334.

40. On Canadian developments, see Colonel K. T. Eddy, "The Canadian Forces and the Operational Level of War," *Canadian Defence Quarterly* 21 (1992), 18-24. Many peacekeeping operations, as well as the army's experience at Oka in 1991, offer useful lessons. On current thinking see Brigadier-General R. A. Dallaire, "The Operational Level of War in the Canadian Forces and the Senior Officer Corps," presentation to the 1993 Senior and General Officer Symposium, Kingston, Ontario, 15 January 1993. Copy in author's possession.

6

Operational Methods of the French Armed Forces, 1945–1970

Sabine Marie Decup

Studies in military history using the term "operational" present particular problems for French scholars. The term has never been given an exact definition in France. Moreover, military historians as a whole disagree about its meaning. In terms of post-1945 French military affairs, some think it is a question of operations—on a relatively large scale—carried out within the North Atlantic Treaty Organization (NATO) area. Others believe that this designation is more narrow, applying to the combined operations performed during the various colonial wars waged by France. The latter view has greater applicability to the French situation. While the documentation concerning the Indochina war is fairly accessible, the same does not apply to Algeria. Due to their sensitive nature, documents associated with the North African crisis are granted to historians via a special dispensation. Accordingly, in dealing with French operational methods in Indochina and Algeria, the best sources remain the books published by the military decision-makers of the time—generals such as Massu and Bigeard and colonels such as Trinquier and Argoud—all of whom had a great impact on their soldiers and influenced the chiefs of staff. This study examines the question of these specific operational experiences from a chronological, rather than a thematic, perspective. This approach is self-evident since operational theories and techniques evolved over time, and because the ever changing political situation, which greatly shaped command actions, must be taken into account.

France emerged from the Second World War weakened by the privation caused by defeat in 1940, the subsequent German occupation, and internal squabbling among its political leaders. Her military capabilities were largely destroyed and her industry nearly nonexistent. She had to reconstruct everything: the armed forces themselves, armaments, and command and control in France and abroad. This was, of course, a hindrance to any external engagement. However, even before Japan's defeat in August 1945, France had to settle the crisis in the Levant and then commit herself in Indochina, where she hoped to reestablish the rights she had enjoyed before the war. For five years, she slowly became enmeshed in this conflict at the far end of the world, into which her best troops were thrown. The nature of the Cold War in the Far East after 1950, with

American involvement in the Korean conflict, led French leaders to hope for a victory in Indochina; but their military doctrine was far from being equal to the task. When, after May 1954, Indochina could be considered lost, the French government suddenly faced the uprising in Algeria. Once again, French forces confronted a guerrilla war. Helped by experience, the army staff instituted new systems and theories better adapted to this form of fighting. But despite operational success on the ground thereafter, the French armed forces were ultimately forced to withdraw due to political factors. The growing importance of the nuclear option was one reason for General Charles De Gaulle's decision to withdraw from Algeria. The nuclear genie brought about a major upheaval in existing strategic ideas. These had repercussions on operational spheres like Algeria and brought French operational theory to a standstill.

INDOCHINA

The difficulties which first appeared in Indochina in 1945 resulted from both the Second World War and the Japanese occupation.[1] Vietnamese nationalism, until then contained by the French administration, burst into the open. Its mentor, Ho Chi Minh, rose as the leader who would rally all the Vietminh forces to fight against both Japan and France.[2] But the arrival of General Jean Leclerc in Saigon on 5 October 1945, as the head of the French task force, challenged Ho. The north of the colony, Tonkin, occupied by Chinese troops, was in a different situation from that of the south, which was occupied by the British and then the French. Although having signed an agreement with the French government, which he argued Paris had broken, Ho rapidly went underground. Thereafter, despite serious problems at home, the French government became preoccupied with Indochina.

The sending of the Second Armored Division to Indochina under General Leclerc was designed to reinstate a French presence. With the failure of the policies of Admiral Thierry d'Argenlieu, the high commissioner, the country was ablaze. After 1946, the fight against the rebels spread throughout the whole territory. Returning to the Far East was not a mere police operation; it soon took on the characteristics of a real war and operational needs increased accordingly.

Political indecision in Paris led France into a new conflict at the same time that it was recovering from the German occupation. De Gaulle's departure from office in December 1946 and the tumult of the National Assembly made the definition of a precise policy more difficult. Without clear political direction, both the civil service and armed forces became more independent, a development that had fatal consequences for the Indochinese policy.[3] The subordination of the military to the civil power after the passing of a statutory order in January 1946 only added to the confusion. The sphere of political activity of the former was therefore considerably reduced. After January 1946, two ministers controlled the armed forces: the minister of the armies and the minister of armament. Every theater of operation had its own commander in chief receiving orders from the government.[4] These commanders had little freedom to maneuver, a state of affairs exacerbated by the ceaseless modifications to the general organization of defense. From 1947 onward, the authority at General Staff Headquarters of National Defense devolved, first, to a committee of the chiefs of staff, then, to a combined staff of the armed forces, and then to a general secretary of national defense having a military rather than a civil nature.

Political instability in France during this period compounded these changes: between 1946 and 1954, 19 governments followed one another whereas during the same period only Ho and his military chief, General Vo Nguyen Giap, commanded Vietnamese political and military forces. Concurrently, the Indochina government had five high commissioners and the French task force, seven commanders in chief. And with a chief of staff in Paris having limited authority, the commanders in chief in the colony did not receive precise orders.

To understand the army's operational choices, the situation must be clarified. Contemporary analyses carried out by field officers are particularly sharp in this respect. More than their opposite numbers in other continental countries, they grasped the global dimension of the Vietminh threat. Only the lack of political instructions prevented them from concluding their analyses logically. At the beginning of the conflict, the French held the towns and most of the vital regions of the colony. On the other hand, their military means remained fairly weak for acting in the northern region, where their adversary took refuge.[5] Because of help from Communist China after October 1949, the Vietminh developed and reached force levels superior to staff expectations.[6] Therefore, the operational side of the war broadened for the French; this was furthered by General Jean de Lattre, the military commander after December 1950, who recognized the importance of the region in the open fight against Communism.

Within this operational evolution, the general staff did not, as a whole, care much about the training of those troops meant for Indochina.[7] New troops destined for the colony had neither specific training for nor any information concerning the events in Indochina. Only those soldiers who returned to France after a first or even a second tour of duty were able to provide some information about the situation. The training provided by such schools as Coëtquidan and Idar-Oberltein[8] only prepared troops *a posteriori* for the style of campaigns conducted in Italy or Alsace! As for the staff, they contented themselves with planning for so-called defensive operations, in reality, mock campaigns meant to confront a hypothetical enemy from the east. In spite of the increasing development of guerrilla war in Indochina, a "modest veil" was cast on these fights as if the outcome were already glimpsed.

Even though the staff in Paris were in a position to put the conflict in its international context, they were not able to determine clearly the conditions of the campaign in the colony. As for the operational chiefs in Indochina, they did not necessarily worry about the political repercussions of their policies; they had to obtain the complete pacification of the countryside, and, to do so, they had to face the imperatives of terrain, their strength and that of the Vietminh, logistics, and so on. Given these considerations, individual initiatives by junior officers became more numerous. Without any strict orders, they had to wage "their war" in order to defeat "their enemy." Still, there remained a few commanders who carried enough weight to forbid any improvised operation. Nonetheless, neither aerial photos, the marking of drop zones, nor air support were used at the beginning of the war. And operational commandos, based on the model of the British Special Air Service (SAS) and among the first troops to have arrived in the colony, only asked to carry out new offensive missions![9]

The operational command also bore the consequences of the changes of government in the metropole, a situation that increased orders and transfers. It was fairly common to see an officer having up to three group commands within 16 months.[10] This was too fast a turnover in a country where a good knowledge of the terrain was indispensable for military efficiency: it reduced the possibility of being evenly matched with Vietminh forces, which were perfectly at ease with the topography of Indochina.

Before 1950, operations aimed at controlling the Mekong delta region and pacifying Cochin China and South Annam. The conquest of the south was fairly successful as a whole, but complete pacification was not achieved: the major communication routes suffered from the Vietminh presence. Indeed, whatever the fighting merit of the task force, its strength was far from sufficient to control the whole territory.[11] In addition, resting on popular resistance, Giap's strategy was too complex for firepower to be decisive by itself. If pacification gradually increased in Cochin China, it still remained extremely superficial and too fragile. Prior to 1950, the command lacked sufficient strength to impose its own fighting method on the Vietminh. Except for the battles of Hoa Binh and Na San, it never held the strategic initiative that could have allowed it to pursue actions at a rate of its own choosing.

The first operations carried out by the task force in South Indochina were meant to free surrounded garrisons, restore the safety of the main roads, and recapture the control of the zones said to be vital or politically important.[12] As mopping-up operations, they made the capture of prisoners and equipment possible but at the cost of high casualty rates. Inexperienced, newly arrived troops would be sent on relief missions. At that time, a lack of resources and limited numbers precluded the combined use of airborne and ground troops. The Vietminh's achievements produced a climate of insecurity,[13] which condemned French troops to passive operations of a purely defensive nature. Constantly dispersed, they were not in a position to play the role to which they aspired: offensive action.

From this perspective, the Indochina war became a "war without a front, like the Spanish war in the past'.[14] The vastness of the territory,[15] the scattering of the vital centers, the method of fighting—new to the French—were elements that, in the long run, tipped the scales in favor of the Vietminh. In 1947, for instance, each battalion had to control a zone of some 6,000 square kilometers with more than 600,000 inhabitants.[16] These difficulties were complicated by fluctuations in political opinion in Paris and Indochina that seemed to vary in proportion to the ebb and flow of military victories.

The year 1947 also marked the time of large-scale operations meant to link the various isolated operational sectors and relieve Hanoi and its hinterland. However, the main targets were rarely hit and the Vietminh escaped. It has to be pointed out that French air transport, overburdened with missions, was largely insufficient to support such large-scale missions. Because of these limitations, and in order to defeat the enemy on his own terrain, it was then decided to use commando forces. Taking part in mopping-up operations, the commandos were often dropped in troublesome zones. Admittedly, they often lacked numerical strength. But in operations to restore order, their actions were generally positive because the ability to resupply by air gradually improved.

The problem of troop levels became crucial in 1948. The Indochina conflict continued, of course, which required additional forces; but there also arose the need to establish a permanent overseas force for the entire French Empire. These two elements nullified any hope of a coherent reorganization of forces in Indochina.[17] For instance, the structure of the airborne troops was far from being adapted to the missions to which they had been entrusted. Similarly, they had not been granted the necessary means. Still, they had to be able to fulfill three types of missions: colonial, European, and in the home country. It was only in the first field that France was more or less in a position to engage well-armed and efficient mobile units.[18] In Europe, she was not in a position to do anything without the help of the Allies.[19] So, in 1948, to strengthen French military power, a reorganization of forces in the home country took place: scattered units were grouped together; their resources were specialized; and new training centers were created. Reorganization also occurred in Indochina, where some units were broken up, command and control simplified, and new bases set up to support operational battalions.[20] In order to meet the enlargement of the operations, an operational staff was created on 13 June 1949, whose efficiency rested on the personality and the value of their officers.

In Tonkin, in order to carry large cleaning-up operations through to a successful conclusion, the high command had to trim defense forces in other regions. Aware of these efforts, the Vietminh undertook political and military acts in those regions to win support from the indigenous population. To reassure the population of the delta, French authorities employed the technique of putting the area under close control: a certain number of watchtowers stretched along the main roads, and companies of soldiers patrolled specific areas. On a permanent state of alert, French forces took turns in responding to intelligence reports about Vietminh movements, helping threatened checkpoints or convoys with direct assistance or, in an emergency, moving against Vietminh command centers, depots, and safe refuges. This continual "stop and go" was entirely different from the big operations in Tonkin.[21] Similarly in Annam, in Cochin China, and in Cambodia, small units of paratroopers were used to exploit intelligence gains or to help regular forces when they were threatened.

The year 1949 was marked by a defensive strategy. While Paris continued to modify the organization of units to make them meet better the operational imperatives in Indochina, the Vietminh suddenly attacked Colonial Route 4 (RC 4). For the first time, the French command realized the military power of the Vietminh. Possessing a strong antiaircraft defense and able to maneuver with great efficiency, Ho's forces showed a capability for going beyond the simple strategy of traditional guerrilla warfare. Within three weeks, the French army lost eight battalions (7,000 men) and a large quantity of equipment.[22] Nevertheless, the Vietminh failed once again: it was not ready to undertake traditional set-piece battles. The strategic and tactical mobility of the French task force, together with its firepower, put a brake on Vietminh operations. As a consequence, Giap decided to wage a war of attrition by having his units infiltrate regions controlled by the French. These units constantly harassed French troops, ambushed them, and began winning the hearts and minds of the bulk of the population.

But the Vietminh could not win set-piece battles. In the delta, being near their bases, French commanders could utilize an array of military assets: armor, amphibious vehicles, artillery, and aviation. Recognizing Vietminh limitations, Giap withdrew into the high and middle regions of Tonkin, where French forces were in a position of inferiority. In the north, far from their bases, the French could only be supplied by air, and they would be unable to bring much of their firepower to bear against the Vietminh.

The RC 4 offensive woke Paris from her drowsiness: it resulted in de Lattre's appointment as commander in chief in Indochina. On his arrival in December 1950, he realized immediately the deterioration of the situation. France no longer faced a mere rebellion, but an effective enemy, well-trained, dedicated, and adept at tactical fighting. To this end, de Lattre established new mobile groups that could rapidly concentrate—with the help of air support and paratroops—to meet Vietminh attacks and take the offensive again. Each command had a paratroop base, which implied an element of air supply. Each operational headquarters had a permanent mission for the systematic study of the physical, human, and tactical characteristics of the action zone of the paratrooper base. All this was supposed to permit the quick and effective engagement of French forces. An increase in manpower, including indigenous soldiers—the "yellowing"[23] of the French army—magnified the operational capabilities of the intervening battalions. De Lattre decided to save the Tonkin delta. The "de Lattre Line" was consequently created to this end,[24] and also to free some troops from the tasks of watching. Thus relieved, these men could reassemble within mobile intervening groups later to become the General Reserve. De Lattre's forces inflicted a major defeat on Giap in June 1951 by repulsing a Vietminh attack along the Day River. After these setbacks, Giap went back to a guerrilla strategy, more adapted to his military situation. He could not venture to go to open country and had to attract the French to the spots where their artillery could not be used. Unexpectedly, stricken with cancer, de Lattre died in January 1952.

By 1952, the war had entered its final phase. In the vastness of the mountainous regions, particularly the Ap Bac of Tonkin, the only asset of the

French task force was its mobility and speed of intervention. Direct support was provided by the air force, including airlift for mobile troops and support forces. Thanks to American help, General Raoul Salan, de Lattre's successor,[25] continued the work of his predecessor and went on with mopping-up operations. But because the Vietminh began a new series of attacks—guerrilla operations were ending—Salan decided to meet them using the troops of the General Reserve under his direct command and with the fortified towns as his bases. After a month's intense work and supplied by airlift, Na San, west of Hanoi, became a true fortified camp;[26] it was capable of defending itself and being used as a support base for local operations. Believing he held an unassailable position, and although all the possibilities that this system could offer remained unexplored, Salan built his whole order of battle on this operational strategy. Thus, the practice of fortified camps became the official doctrine of the French task force, and the evacuation of the small posts in Laos was undertaken. The Communist offensive fell foul of French strongholds in the Plain of Jars and Luang Prabang. The French achieved this defensive success thanks to excellent intelligence collected by the pro-French Vietnamese *maquis*, the swift intervention of light units, and, despite some limitations, air support.

General Henri Navarre succeeded Salan in May 1953; he also wanted to break the Vietminh offensive and wage a "running war."[27] Mopping up operations on a large scale continued, but, by the summer of 1953, the Vietminh battle corps was fully restored and ready for a general offensive. On the model of the bases and fortified camps, and after realizing that the Vietminh's target was the invasion of Laos, Navarre launched Operation Beaver to recapture Diên Biên Phu. Diên Biên Phu lay astride the lines of communication between Hanoi and Laos. A fortified camp was established around a strongly defended airfield. The base was set up as the hub for logistical support for the mobile units, artillery, and warplanes that would radiate operations outward to confront the enemy.[28] Long-distance safety for the latter three elements of the French forces was to be augmented by the *maquis.* However, this concept of a resisting center to wear out enemy forces was criticized from the end of 1953. French forces in the region were weakened by the need to divert troops to the delta. In addition, there could be no guarantee of adequate air and land transport. Scattered and sheltered, Vietminh artillery could not be hit effectively.[29] As a resisting center, Diên Biên Phu could not be held for long.

The Vietminh attacked on 13 March 1954, and quickly demonstrated the superiority of their artillery, anti-aircraft defenses, and logistical capacity. Within days the French garrison was under seige, their land routes cut and the runway open to enemy fire. Giap's forces launched a general assault at the beginning of May, an action that sounded the death knell of France's presence in Asia.

To understand French operational methods in Indochina between 1945 and 1954, it is important to appreciate the constraints placed on French forces by their equipment and manpower. Of course, the failure of pacification policies in the late 1940s was due to an inadequate command and control system and an underestimation of the revolutionary fervor of Giap's men. To avoid a protracted war, a great deal of operational activity was necessary. Such activity was dramatically lacking because of the need to have limited numbers of troops in both the delta and in the Ap Bac. And the incapacity of the politicians and the mistakes of theater commanders also helped prevent victory. But the inadequacies of their equipment was also responsible for French ineffectiveness. The material means available were insufficient to meet military needs, and the equipment was often old. For instance, warplanes (Junker-52s and Spitfires) provided by the British during the Second World War and afterward were insufficient to the needs of a war in which airpower might play a prominent part.[30] These planes wore out quickly and were difficult to maintain because of the lack of spare parts. French commanders could only make an "economical war":[31] limited arms and supplies put a brake on commanders' ambitions and the conduct of an offensive war.

Considering the choice of combined operations—250 in nine years—insufficient means restricted the objectives. Taking advantage of this situation, the Vietminh deployed troops widely and attacked many fronts at once. Forced to support land maneuvers, the air force was considered as a secondary arm. Its role of watching and reconnaissance soon proved wasteful, considering enemy mobility. When, at long last, it was entrusted with support missions, its fighters were too fast and too poorly adapted for antiguerrilla warfare. Transport aircraft bore the brunt of the war, given the absence of passable roads. As for the helicopter, it was hardly used in Indochina. The few missions carried out at the end of the war were for medical evacuations, this at a time when the Americans began to realize the tactical offensive capability that this aircraft could offer.

All the commanders in chief agreed about the need for training a local army. Such a force would have been entrusted with static defense, leaving to the French task force and paratroopers to undertake combined interventions.[32] In this way, a general reserve of forces could have been created, capable of acting anywhere in a minimum amount of time and allowing the reserve to take the offensive. But the command, albeit well informed, did not have the necessary means to achieve the different types of operations required to ensure victory. The protection of the pacified regions, the mopping up of threatened areas, and the destruction of the Vietminh battle corps demanded too many men. Organizing pacified villages or those able to ensure their own self-defense was not enough to rally these populations completely.[33] For example, the self-defence group was not sufficient to pose a real threat to the Vietminh. Additionally, the operational use of local forces, led by a few French officers, had mixed results.

After 1950, the French task force numbered about 190,000 men divided into 90 battalions. Part of this force was reassembled into mobile units comprised of three infantry battalions, three to four paratroop battalions, and an artillery group. Sometimes two armored groups, plus two amphibious groups, were joined to these mobile units. Adapting the equipment to the terrain remained of a paramount importance. A new generation of amphibious vehicles—Crabs, then

Alligators—were utilized; they could move on firm ground or in the desert, as well as on flooded lands. They were ideally suited to operations in the delta; and they were also able to provide fire support, collect and transport parachutes, and take on the wounded. It goes without saying that, without them, no major operation could have been possible in the Mekong delta.[34]

As for the air force, some 250 fighter aircraft[35] were regularly thrown into the battle from 1950 onwards. There was also a reserve of airplanes for reconnaissance. Once the air force's prominent part in combined operations had been recognized, its orders were defined: fire support, transport, and information under the form of direct or independent backing actions. It carried out road reconnaissance and bombing operations, intervened in armed engagements, and parachuted reinforcements. Thus, airpower mainly supported the actions of the army.[36] Its bombing targets were vital elements of the economic infrastructure such as dams, dikes (destroying Vietminh rice production), and supply depots. Realizing the importance of the economic stakes and lacking planes of their own, the Vietminh established strong anti-aircraft defenses.

Apart from its traditional missions like the struggle against arms traffic, the navy took an active part in inter service missions. Its specific character was important for protecting river transportation routes, which were particularly numerous.[37] Naval warships often provided artillery support, a role which increased over time because of the mobility and relative invulnerability of warships—Vietminh forces lacked a maritime component. The warships' "shooting against land" supported the major operations carried out along the coasts and rivers when the latter were deep enough. French gunboats helped in the pacification of the deltas and in mopping-up operations there. They destroyed junks and sampans that carried equipment to Vietminh bases and allowed river convoys to sail safely.[38] Naval air took an active part in the conflict: some were even employed at Diên Biên Phu.

A top priority for the high command, the delta was surrounded by a belt of concrete fortifications built on its periphery. They were designed to permit the defenders to resist the invader until reinforcements arrived, namely the mobile intervening corps. In this region, pacification was achieved by the operational mobile administrative groups. Their mission was to protect the emerging economic, social and health infrastructure by shielding the "mopped up" areas from Vietminh influence.[39] If non-Communist Vietnamese leaders—civil servants, administrators, teachers, and doctors—helped to achieve economic and social stability, the pacification process dear to old colonials might be able to survive without the mobile intervening troops.

The notion of the General Reserve appeared in Indochina. A certain number of battalions were put at the commander in chief's disposal, the latter being thus able to deploy them depending on need. For instance, in 1949 Tonkin had its own paratroop battalions and, therefore, the Saigon General Reserve was only meant for intervention in Cochin China, Annam, or Cambodia. As a whole, one company was put on the immediate alert, ready to board; a second was on 12 hours' alert; and a third on 24 hours' alert. They were to undertake small operations in the immediate vicinity of their base, and, each week, the companies swapped roles.[40] This structure was not applied to other battalions which lacked suitable lift for rapidly mounted operations. The French task force had vital elements: armored vehicles, artillery, aviation, and elite corps such as paratroopers and *Dinassauts*, tiny naval "task forces" that were particularly efficient for controlling the rivers. Their massive use forced Giap to choose subversive action over pitched battles. The example of the *Dinassauts* is instructive. Their flexibility enabled them to adapt more easily to subversive war along the important water routes. And they participated in landing troops and in other big operations.

The Indochina *maquis* also contained "action service" troops entrusted to undertake antiguerrilla operations against the enemy defense system. They had some success in rallying the population and holding Giap's men in check. Belonging to the General Reserve, they were in a position to contribute to the pacification of some areas. This element of the "action service," called the Group of Airborne Mixed Commandos,[41] based its operations on intelligence received from the External Defense and Counter Espionage Service. It had the same basic organization as a troop corps and could be used to support the main Action Service.[42] Its mission also consisted in creating other *maquis* units, training sabotage teams, and setting up escape networks;[43] they also participated in planning airborne operations. The arrival of intervention commandos was therefore better prepared and indeed more discreet, which enabled them to succeed in launching quick raids, deep into the country, keeping it in a state of insecurity for the rebels. The *maquis* were entrusted with a true mission of an operational nature: "prevent the population from collaborating with the Vietminh—create a permanent climate of insecurity in the Vietminh rear—gradually bring the population to participate in the action by a systematic destruction of the political and military organization set up by the Vietminh in the rear."[44]

Yet, confronted by Giap's well-trained units and although of equal manpower, French troops did not measure up to the situation. The Vietminh gradually succeeded in organizing a powerful land battle corps, superior to French mobile forces. Giap's eight divisions were seasoned, more effective in using the terrain to their advantage, and, ultimately, better equipped thanks to China's help. They also had effective artillery and a not inconsiderable antiaircraft defense. In spite of some aspects which were original in conception and execution, the operational elements of French military action in Indochina remained rather classical—corresponding to the model of the Second World War. They were not really adapted to a new situation. At the Geneva Conference of 1954, which began as the battle of Diên Biên Phu entered its final stage, the colony of Indochina fell from French control. Laos and Cambodia achieved independence. Ho Chi Minh emerged as the leader of the Democratic Republic of Vietnam, centered in Tonkin. The south became the Republic of Vietnam, which found support from the United States. France's moment as a Far Eastern power had passed.

ALGERIA

After the humiliation suffered at Diên Biên Phu and the Geneva Conference, the French army was almost at once deployed in Algeria. North Africa then became the new battlefield against Communism, an ideological struggle which, importantly, was recognized by France's NATO allies. Very quickly, however, the French command was again confronted with an ambiguous mission. Troops in Indochina were transferred straight to Algeria, and a rather cumbersome military system was set up while the North African rebellion was still in its infancy. With Indochina veterans suddenly representing 80 percent of the French contingent in Algeria, the operational approach to this conflict was influenced by the recent East Asian experience. Just as important, given Algeria's constitutional union with metropolitan France, it became a conflict which was increasingly affected by domestic French public opinion.

With rebel attacks initially having little effect, the colonial government misjudged how to meet them. As for the first commanders in chief, they failed to understand the nature of the rebel action, which, to say the least, was not very logical. However, in spite of a few French military successes, the situation deteriorated. Tempted to apply the methods used against the Vietminh straightaway, the high command did not immediately realize the differences between the two campaigns. Not only was the terrain entirely different but so, too, was the enemy: both its way of living and its philosophy. By 1955, the Algerian guerrilla movement, the National Liberation Front (FLN), was winning widespread support and increasing its terrorist attacks. It was also a period when international awareness of the scope of this new conflict was spreading.

Initial operations by French units were almost always independent of one another. The decision to have a unified command was taken only in 1955. Soon after, contrary to the delay experienced in Indochina, an airborne reserve was formed capable of intervening in Morocco and Tunisia, where disturbances were also likely to arise. The use of troops and the general methods of their training were studied while air supply and deployment of units was defined. FLN terrorism heightened in mid-1956, just as the French troops were being readied for the Suez intervention. As this story is well known, there is no need to dwell on it. But once back from Egypt, French troops were widely dispersed in Algeria to face the terrorist threat that had been amplified by the political defeat suffered by France and the United Kingdom in Suez.

Paradoxically, the French army came out of the Indochina war as a capable instrument of anti-guerrilla warfare. The requirements of revolutionary war as far as targets, means, and methods are concerned were well known. Still, the army's first steps were unsteady. Its first reaction was to use techniques established by nineteenth-century generals like Bugeaud and Lyautey, who had conquered Algeria. Such was the case for covering operations, which were being organized as early as 1956. But, since the struggle resulted in heavy casualties, reinforcements were dispatched and revised tactics employed: forces were used to search difficult regions and twinned with surface control designed to strengthen the defense of roads and weak points. Consequently, reinforced control areas, even in isolated areas, were created to improve security; in some places, traffic and parking were forbidden.[45] The regions in turn were divided into operational zones depending on geography and other factors. In 1958, the command concentrated the rural population near posts held by the covering troops. These were generally set up in farms temporarily abandoned by the French settlers at the beginning of the troubles. The aim of these concentrations was twofold: to prevent the FLN from influencing the population or taking reprisals against those Algerians supporting the French; and to allow the army to undertake administrative, medical and educational reforms.[46]

Therefore, the army became increasingly responsible for nonmilitary tasks which blunted its effectiveness. Despite large-scale maneuvers, rebels would escape. Fighting techniques were ill adapted to a form of guerrilla threat different from that of Indochina; moreover, the army had to take on political responsibilities which came to a climax when General Jacques Massu was granted emergency powers on 17 March 1956. Deprived of clear directives from Paris, the army was bound to be implicated in Algerian political life: finding it difficult to be neutral, it gradually defined a political and social philosophy of its own.

Before this, in 1955, it did become clear to the colonial government that means had to be found to avoid the waste of men and equipment. This recognition produced a reorganization of the army that saw it adapt operational methods used in Indochina to the unique Algerian terrain. In addition, there was an added emphasis on intelligence; in 1956, the Interarm Coordination Center, a permanent intelligence infrastructure, was set up to link operational units with one another as well as with the territorial commands. Its mission was to keep watch on the armed bands and to spot the underground activities of the FLN. Its basic units were called "operational detachments of protection" (DOPs), and each had an officer in charge. Their staff comprised military policemen, interpreters, and a secretarial and maintenance staff.[47] The DOPs were to coordinate and guide the actions of the intelligence officers inside operational units. They proved critical in the regions where the rebels had a political and terrorist infrastructure.

The war also triggered a reorganization of the French armed forces outside Algeria, since a large number of active and general reserve army units stationed in the home country and in Germany were transferred to North Africa to the detriment of NATO. However, heavy equipment was often left behind, for instance, AMX tanks, which were totally ill suited for fighting guerrillas in the *djebels*. However, during the Indochina war, the use of pro-French elements among the indigenous population proved particularly effective: this was the reasoning behind the use of Moslem commandos in Algeria. Although these natives proved to be positive assets against the FLN because of their knowledge of the country and its customs, there was room for improvement and they also had to be trained.

Because of its development to work within NATO, the air force was ill prepared at first for its mission in North Africa. Both aircraft and their support infrastructure were inadequate for anti-guerrilla operations. To fill the gap, the Light Aviation of the Army (ALAT) was created as early as November 1954. The intensification of military actions in Algeria brought about its expansion as Reserve Flights of Support Light Aviation (ERALA) and development on reconnaissance missions.[48] Shaped by its function, ERALA fitted to the originality of its mission throughout the war. This flexibility permitted airpower to play a decisive part in the struggle against the FLN.

To acquire this flexibility, the air force organization was modeled on the army. A Tactical Air Group comprising three units was created in 1956; these corresponded to the three military regions in Algeria—Constantine, Oran, and Algiers. Through this means, a geographical decentralization of airpower took place: flexible, quick interventions became possible. It was all the more workable as, corresponding to ground operational sectors, each region was divided into air command posts (PCA) with substantial transmission equipment. From 1957 onward, the army corps of each region had a tactical support group, which included helicopters, to undertake liaison, reconnaissance, and land support. Each air command was divided into four operational areas.[49] Lastly, whereas in Indochina there were but few helicopters, all used for medical evacuation, in Algeria these airships became operationally important. Helicopters totally changed the dynamics of battle by freeing the infantry from the constraints of the terrain and reducing strain on troops by carrying them to the operational spots.[50] Although first conceived of as a means of transport, helicopters passed with hardly any transition to assault aircraft; hence, heavy helicopters like the H.34 (commonly called Sikorsky 58) were used for transport as well as for assault and fire support.[51]

Finally, the French navy also played a prominent role in combating the FLN, though it could not do as much as it had in Indochina because of the desert conditions of Algeria. Its chief task was interdiction: providing an efficient means of preventing the rebels from receiving waterborne supplies. The naval headquarters at Mers-el-Kébir (SURMAR) controlled the 1,300 kilometers of coast in Algeria, from the innumerable inlets to the open sea. To do so, it used small and large patrol boats, dredgers, and escort ships, while the fleet safeguarded the western Mediterranean.[52]

When the question of teaching the new methods of anti-guerrilla warfare arose, the government asked Colonel Marcel Bigeard to set up a school for the training of young Staff School officers. At Jeanne d'Arc, near Philippeville, he organized instruction in non-conventional warfare. One hundred and twenty captains received lectures from instructors, who included many ex-servicemen with experience in Indochina and Algeria. Practical training—maneuvers with operational units and parachute jumps—was also organized. The goal of paratroop training was to have the graduates in peak physical condition.

Explaining and solving concrete problems in simulated battles were especially beneficial since these officers were generally more expert in the theories of warfare. Bigeard's goal was to strengthen their morale so that they could perform more efficiently in the field. But beyond this aim, these officers were prepared at the end of their training for the unique circumstances awaiting them in Algeria, especially at commanding via radio and in arranging combined ground and air operations with infantry weapons, artillery, and air support.[53]

After the Suez crisis, rebel groups were regularly receiving weapons from Tunisia and Morocco.[54] FLN units were stationed outside of Algerian territory, where they also trained, rested, and refitted. The struggle against the rebellion became more onerous with the intensification of terrorist acts. Salan understood the situation when he assumed the post of commander in chief in Algeria on 14 December 1956: French forces had neutralized only a small proportion of the rebels. Accordingly, barbed wire barrages were built in crucial locations, together with electrified networks. The 300 kilometer Morice Line and border barrages reduced 70 percent of FLN supplies, which freed regular units and made possible their operations within the General Reserve inside the country. Paratrooper regiments received the task of intercepting the remaining groups in the rear of the barrage. Posted along main routes, they were occasionally reinforced by the sector units. All investigative means were deployed to expose enemy positions and, as soon as a group was located, all fighting and support means were concentrated in the area via helicopter transport. Anti-revolutionary warfare had three constituent elements in North Africa: the destruction of FLN bands in Algerian territory, the initiation of political and psychological actions leading to the rejection of the rebel movement by the indigenous population, and the interdiction of rebel forces and arms into the colony. The third of these was achieved as the border became safe; by 1959, Algeria was a vast combat area.

The conduct of military operations differed for those French troops deployed in urban areas. They could be used to oppose riots and mass demonstrations, but their role was mainly limited to patrols, ambushes, and terrain searches in hostile neighborhoods. On the other hand, the police played a prominent part in combating urban terrorism. But, initially, whereas the police were well trained to operate in the cities but lacked means, the army had the means but not the training. Still, it was fairly simple for the army to convert to urban warfare. It had to consider this struggle against FLN underground networks of action and propaganda as a regular operation, one against an armed and organized enemy. However, urban terrorism was not combated by force deployments. Instead the operation was largely psychological, aimed at putting the population's mind at ease and creating an insecure climate for the rebels. The army's real work focused on exploiting intelligence gains to locate and destroy the terrorist networks.[55] French intelligence used various means to ascertain the nature and size of FLN terrorist cells.

At the beginning of 1957, when the first Bigeard troops arrived in Algiers, their mission consisted of nothing but a show of force to deter the rebels.

However, they quickly grasped that the return to calm and security could only be obtained by destroying the terrorist organization. As they could rely on neither the existing military elements in the city, which were short of means, nor the police, which had limited intelligence resources, they built a new operational structure. Determining the framework of the FLN organization became the first target. All intelligence had to be rapidly exploited to permit infiltration—the FLN was divided into sectors, groups, cells, action branches, and political branches. The paratroopers concentrated on the destruction of bomb-producing and armed groups. But these cells did not function only within urban boundaries; they had extra-urban branches, which meant that they had to be hunted down far into the *djebels*.

General Maurice Challe succeeded Salan as commander in Chief after De Gaulle returned to power in May 1958. The government in Paris had collapsed because of a domestic crisis caused by the political and manpower costs of the Algerian war. Moreover, Salan and other hard-line senior officers were openly dabbling in Algerian politics and, to aid in finding a settlement, De Gaulle engineered this change in command. Under Challe, combined operations expanded in scope. Through these missions, carefully planned and conducted by the operational groups, the army's mission was to neutralize the rebel presence throughout Algeria, from the Sahara to the *djebels*. In this process, French forces sought to divide FLN units into smaller groups to better suit the fighter commandos charged with controlling each sector. Moving from west to east, French operations followed the different steps of pacification. Thanks to the superiority of the French armed forces, which outnumbered the FLN, rebel bands were surrounded within medium distance and closed into a perimeter that, progressively, became narrower. Once the sealing off was achieved, elite forces, supported by helicopter-borne reserve units, would clean out the area.

In December 1958 important operations brought into play units of the General Reserve. But the FLN could not be broken. Confronted by this half failure, Challe decided to establish units whose aim was to pursue rebel bands inside the *djebels*, create psychological insecurity within the FLN, and, from this, destroy FLN military potential. Backed by sector intervention units and the air force, special light units were charged with enforcing thorough pacification. When this strategy proved insufficient, for example, in mountainous areas, reserve units were also used.

Favoring night time deployments, Bigeard was a specialist in this kind of large-scale operation. His units were often on the alert: when intelligence was received, some would be driven by trucks to selected points; others would be waiting, ready to be transported by helicopter if reinforcements were needed. Along with helicopters, fighter air support and artillery were always available. Once the troops had been landed, they would deploy to take stock of the situation at dawn. By radio and, sometimes, by directing operations from a command helicopter, Bigeard would coordinate helicopters and transports to surround the rebels.[56] All of this implied a unique style of command, whose network of VHF radios enabled all French forces involved to follow the maneuvers.[57] It was through the impetus given by Bigeard and others that the classical conception of guerrilla warfare was modified: an interservice operational tool, flexible and mobile, was always on the alert. Intelligence was the cornerstone of the whole system. Rebel troops or their depots could not be spotted without it. Air-land cooperation permitted French forces to win the operational side of the struggle.

The French armed forces won the military side of the Algerian war; but they had little control over its political dimension. De Gaulle's rise to power in 1958 spelled the end of French control over Algeria; wanting to protect France's domestic stability and prevent the undermining of its international position as a great power, he was prepared to abandon the colony. Thus, he decided to allow the Algerian people to determine whether they wanted to continue their political connection with France or choose independence. Although it took three years to arrange a plebiscite on such lines—and after having to contend with a potential revolt by the army leadership in Algeria (which included Salan, Challe, and Massu)—the Algerian people chose independence. This time the loss of an integral part of the French Empire could not be blamed on the army.

OPERATIONAL POLICY IN FRANCE'S NUCLEAR AGE

The discussion of the post-1945 nuclear era will be shorter for two reasons: first, its theories have yet to be applied; and, second, the operational plans of action largely remain secret.

When De Gaulle returned to power in 1958, he redefined France's position in NATO. The end of the Algerian war and the demobilization of large numbers of regular forces increased his room to maneuver. The rebuilding of the French armed forces, free from colonial burdens and organized around the nuclear weapon, gave him the political and military means to achieve his ambitions within NATO.

Between 1954 and 1956, because of the Algerian war, France removed three divisions from Germany and, therefore, deprived NATO of some of its best units in Central Europe. Throughout the 1950s, the operational role and mission of the French land and air forces decreased in Central Europe. French disengagement from this military structure was slow; but it was carried out while her territory and her airspace contributed to the common defense, especially since her army and air force were to reinforce those of the Americans and British.[58] While her operational mission was reduced, her logistical one was enhanced because of France's geographical position. Indeed, the main American lines of communication passed through French territory, and some important elements of NATO's armed forces were also stationed there.

But over time and within a global perspective, France's membership in NATO was viewed as incompatible with the responsibilities that De Gaulle and his advisers wanted to assume outside Europe. The French contribution to NATO's common defense implied the rebuilding of a coherent army so that she could maintain her rank and her objectives.[59] De Gaulle's theory from 1966 onward was founded on two concepts: a disinclination to participate in making NATO's military policy and autonomy in foreign policy decision-making. From the perspective of overall strategy, nuclear power was a political instrument. That is why there evolved in France the reduction of conventional forces, the simplification of their structure, and the progressive concentration of military decision-making in the hands of the president of the Republic.[60]

In terms of the use of nuclear weapons, the "operational" level did not exist in itself; it was only in the case of the practical application of strategic and tactical theories that it could have been defined. Nonetheless, French doctrine combined three notions: a strategic reserve of the armed forces to be held back for the decisive battle; a struggle combining nuclear, classical, and guerrilla warfare; and a long war executed by new forces, those of the operational Defense of the Territory (DOT). That is why France was, at that time, developing offensive nuclear forces as well as conventional troops. In the field of air-power—before the advent of ballistic missiles—this saw the development of the "Mirage" warplane, designed both to carry a nuclear bomb and act as a fighter. Then, as Algeria demonstrated, the helicopter became important because of its operational utility. The structures and the missions of the air force were significantly transformed to help it become the main element of the new French strategy.

To facilitate the double mission of the armed forces, the navy also received new weapons: the aircraft carrier and the submarine—France's first nuclear submarine was launched on 29 March 1967. At the beginning of the 1960s, the navy was still the privileged arm of any overseas action, despite the advent of long-range transport aircraft. All overseas operations to a nonlandlocked country supposed the use of an amphibious force augmented by the fighting and technical resources of the French navy. The withdrawal of French Mediterranean naval forces from the NATO command took place on 7 March 1959; it was the first step in the process of disengagement from the integrated military structures. In 1960, France refused full participation in NATO's integrated air defense system.[61] Little by little, and within the context of a graduated response, the growth of the Federal Republic of Germany's Bundeswehr and the implementation of a "forward defence" along the Iron curtain designated the French forces as an operational reserve within NATO. Finally, in 1966 France withdrew from the NATO command.

A turning point for France's operational theories occurred in 1967. This involved the decision to use French conventional forces to intervene in trouble spots abroad judged crucial to France's national interests. The idea was that crises liable to develop abroad—primarily in former French African colonies—called for rapid action of a limited scope rather than massive intervention, as had been the case in Suez.[62] Maintaining links with her colonies and former colonies was one of the chief pillars of France's post-Algeria global policies. In the 1960s, France had imperial possessions in the Pacific Ocean (for instance, Tahiti) as well as in the Caribbean (for instance, Martinique); and because she opted for cooperative links with those African colonies which received independence after the Algerian war (such as Senegal), she possessed a unique position in that continent. This cooperation was born from a growing number of technical assistance agreements with the newly independent states.[63] France occupied several bases in these countries and was willing to defend common economic, political, and strategic interests. Part of her forces were permanently stationed overseas to be in a position to assist French intervention forces if the need arose. All this came within the framework of a deterrent policy: to be credible, this policy demanded a show of resolve to act each time an essential interest was at stake. What distinguished these new interventions from earlier ones such as Suez was the request of a sovereign government to France to send her soldiers to operate in their country. Most of the time, the relationship with the African countries remained, which induced them to ask for France's help during emergencies.

These overseas interventions required professional forces, being at the same time well trained, motivated, and ready to be sent abroad at short notice. Elements of these forces had to be pre-positioned, while others, within the metropolitan army, had to be on constant alert, ready to be dispatched by available maritime and air transport. Command centralization was as much justified by the dominant role played by the president of the Republic in foreign and defense policy-making, as by the planning involved for effective intervention —for instance, permission to establish overseas bases on the territory of other sovereign powers. Added to all this was the necessity of having detection and long-range transmission devices available and, later, satellites and radar, forward based aircraft and other resources, without which no remote battle could be fought.[64]

In February 1960, one month after acquiring its independence, Cameroon called for France's help to curb internal disturbances. French forces isolated the zone where the rebellion prevailed and crushed it *manu militari*. In February 1964, French forces intervened in Gabon. Leon M'Ba, the president, was the victim of a putsch and had been taken prisoner by the rebels. Thanks to the intervention of French troops, transported by helicopter, M'Ba was freed and restored to his post.[65] From April 1969 to December 1970, several units took part in the pacification undertaken in Chad, just as they had participated in that of the Somali Coasts in 1967. Actually, in 1968, De Gaulle answered a request from President François Tombalbaye—although the latter was known for disreputable methods of government—and assented to the involvement of the French army to help suppress a rebellion in the north and east of the country. French units thrown into the battle achieved considerable success, and the rebels suffered serious losses both in men and armament in spite of the poor conditions of climate and terrain. The Foreign Legion distinguished itself, but nothing could have been achieved without the air force's numerous hours in flight. The missions of the air force drew upon the lessons of previous anti-guerrilla struggles: reconnaissance, fire support, liaison between indigenous and French troops fighting on the ground, and helicopter transport or medical evacuation.[66] In reality, this intervention was motivated for the same reasons as that of the nineteenth-century colonial conquest: to secure an area of paramount strategic importance. This is the keystone of the French presence in the Sahara region, which is still subject to Arab-Moslem resistance.[67]

French conventional forces did not only operate overseas. As early as 1961, they participated in army exercises with contiguous countries, designed to harmonize current doctrines and operational methods. The first example, Operation Colibri, took place in Germany in 1962. In 1964, the first exchange scheme with the Spanish army occurred when a helicopter exercise was performed. Yet, even this experience was not completely satisfactory because it was too limited.

CONCLUSION

The absence of political-military directives during the Indochina war was one of the factors that impaired French operational development in that crisis. Lacking adequate means and command structures to implement effective operational methods, the French army was too rigid and unable to adjust to varied circumstances. This prevented it from successfully undertaking anti-guerrilla warfare in Southeast Asia. On the contrary, in Algeria the staff came to understand how to meet guerrilla actions, this by gradually adopting the notions of the General Reserve and the combined use of arms, in which the helicopter was of prime importance. It turned out that collaboration between the army and the air force was the only efficient means to conduct anti-guerrilla warfare. The French forces, and mainly its staff, then used these methods in their African operations in the 1960s and into the 1970s. For this purpose, they developed interservice cooperation inside an army—a professional one in such a case—using operational methods without really knowing it. The introduction of nuclear weapons to the French national armory in the 1960s implied significant operational changes: within a few years, the army passed from the big battalions of the First and Second World Wars to that with a more technical orientation. The Gaullist period was favorable to a harmonious union of French foreign and defense policy: at last, there existed a coordination between diplomacy and military operations.

In the years following the period covered by this study, the links between France and her former colonies have loosened: Some agreements were revised (that with Senegal); others were denounced (those with Mauritania, Madagascar, and Niger). This induced the staff to set up an intervention force in the home country and to reduce the permanent presence overseas. It appears that France has had effective operational means for about 30 years without employing such a term to describe the theory and practice of its military policy. This concept of intervention, thus, now refers principally to putting out "brush fires" likely to destabilize the country in which they break out[68] and which threaten the global balance of power, or narrower French strategic interests.

NOTES

1. As early as 1940, because of the accommodating attitude of the Vichy government, Indochina came under Japanese occupation.

2. On 2 September 1945, he proclaimed the independence of the "Democratic Republic of Vietnam."

3. J. P. Rioux, *La France de la Quatrième République*, vol. 1 (Paris, 1980), 138.

4. A. Martel, *Histoire Militaire de la France*, vol. 4 (Paris, 1994), 243.

5. The latter created some "liberated zones" in regions difficult to access. There he had a de facto power, introduced the sharing out of land, and controlled the economy, schools and police.

6. The arrival of the Chinese troops at the frontier with Tonkin allowed the Vietminh to be resupplied with arms and food, and to find there bases for training and resting.

7. M. Schmitt, *De Diên Biên Phu à Koweit City* (Paris, 1992), 21-22.

8. Artillery school.

9. P. Sergent, *Paras-Légion, le 2ème BEP en Indochine* (Paris, 1982), 26.

10. It was the case for General Maurice Schmitt before he left for Diên Biên Phu.

11. P. Franchini, *Les Guerres d'Indochine*, vol. 1 (Paris, 1988).

12. Martel, *Histoire Militaire*, vol. 4, 296.

13. They sent their troops on pinprick missions to oblige the French to multiply their missions.

14. Schmitt, *Diên Biên Phu*, 24.

15. The land is nothing but contrasts: The flat delta of the Red River—where the vast majority of the population was concentrated—is opposed by the covered jungle of the high region.

16. Franchini, *Guerres d'Indochine*, vol. 1, 375.

17. P. Gaujac, *Histoire des Parachutistes Français*, vol. 1 (Paris, 1975), 164.

18. Light units, like infantry battalions, generally had armored jeeps and means of radio transmission, but no artillery at their disposal.

19. She had neither gliders nor transport aircraft and was not sufficiently organized.

20. They assumed the logistical and technical aspects of the fighting battalions, were in charge of resupplying by air, and had to maintain the efficiency of a coherent warning system.

21. Gaujac, *Parachutistes Français*, vol. 1, 196.

22. Thirteen guns, 450 vehicles, 120 mortars, 3 armored platoons, 2,140 machine guns, 8,500 rifles; see Gaujac, *Parachutistes Français*, vol. 1, 245.

23. In most cases, it reached 50 percent of the strength and allowed a better adaptation to the terrain.

24. A line of fortified towns around the delta.

25. Salan was appointed on 6 January 1952.

26. The camp had 12,000 well-equipped and well-trained soldiers and was surrounded by 1,100 tons of barbed wire and 3,500 mines.

27. Gaujac, *Parachutistes Français*, vol. 2, 376.

28. Martel, *Histoire Militaire*, 310.

29. Schmitt, *Diên Biên Phu*, 71.

30. At the beginning of the conflict, only one-third of French aircraft were operational: 25 out of 45 transport aircraft, 18 out of 30 Spitfires, and 30 out of 50 Morane.

31. Martel, *Histoire Militaire*, 300.

32. M. Carver, *War Since 1945* (London, 1990), 119.

33. Ibid., 377.

34. R. Gaget, *Commandos Parachutistes* (Paris, 1992), 98.

35. This number includes naval aircraft like the Bearcat and Helldiver fighter-bombers, the B-26 Invader and Privateer bombers, and the Dakota C-47 and C-119 transport aircraft.

36. A. Zervoudakis, "L'Emploi de l'Armée de l'Air en lndochine, 1951-1952," *Revue Historique des Armées*, 186 (1992), 87.

37. There were approximately 41,000 kilometers of rivers and 3,000 kilometers of canals.

38. M. Battesti, "La Marine et la Guerre d'Indochine," *Revue Historique des Armées* 177 (1989), 85.

39. Martel, *Histoire Militaire*, vol. 4, 308.

40. Sergent, *Paras-Légion*, 25.

41. On 1 December 1953, the GCMA changed its name and became the GMI, the Mixed Group of Intervention; after this date, it was no longer an airborne unit.

42. This did not prevent it from acting at the time of the departure from Na San; it allowed the task force a better execution of the Navarre Plan.

43. R. Tringuier, *Les Maquis d'Indochine* (Paris, 1976), 39.

44. Ibid., 96.

45. J. M. Marill, "L'Héritage Indochinois: Adaptation de l'Armée Française en Algérie, 1954-1956," *Revue Historique des Armées,* 187 (1992), 29.

46. From June 1958 onward, 32,000 people were grouped within 22 centers.

47. H. Le Mire, *Histoire Militaire de la Guerre d'Algérie* (Paris, 1982), 67.

48. The first light aircraft squadrons, created in June 1955, were reorganized in March 1956 into the Group of Light Support Aircraft. See "L'Armée de l'Air en Algérie: La Lutte Anti-Guerilla," *Toute L'Aviation* 8 (1992), 217.

49. Gaujac, *Parachutistes Français*, vol. 2, 485.

50. Le Mire, *Guerre d'Algérie*, 49.

51. For additional information on the use of helicopters and their mission, see SHAA, I 157, 1958: Règles de l'emploi de l'aviation de renseignement et d'appui feu.

52. Le Mire, *Guerre d'Algérie*, 91.

53. Using the Piper, they could guide fighter patrols and adjust artillery fire.

54. Beginning in 1957, 300 weapons per month were received from Morocco and 500 from Tunisia.

55. M. Bigeard, *Pour une Parcelle de Gloire* (Paris, 1975), 280.

56. On the essential characteristics of surprise, speed, mass action, and flexibility, see SHAA, I 157, 20 février 1958, no. 5OO/EMAA/3/OP/DR: Instruction provisoire d'emploi des hélicoptères en AFN.

57. "L'Hélicoptère Lourd dans les Opérations de Maintien de l'Ordre en Algérie," *Forces Armées Françaises,* 150 (1959).

58. F. Bozo, *La France et l'OTAN*, (Paris, 1991), 55.

59. Ibid., 66.

60. J. Doise and M. Vaïsse, *Diplomatie et Outil Militaire 1871-1991* (Paris, 1992), 609.

61. A compromise was made that put the French alert warning systems and part of the forces from the north under allied command. Although this agreement infers that they were integrated, the order to fire was given by a national military authority.

62. In this regard, the air force was given the first Transall C-160 and the following year the DC-8, an intercontinental four-engined aircraft. A. Foures, *Au-Delà du Sanctuaire* (Paris, 1986), 40.

63. P. Dabezies, "Les Interventions Françaises Outre-Mer," in D. David, ed., *La Politique de Défense de la France, Textes et Documents* (Paris, 1989), 158.

64. Ibid., 160.

65. P. Biarnès, *Les Français en Afrique Noire de Richelieu à Mitterand* (Paris, 1987), 367.

66. Foures, *Au-Delà Sanctuaire*, 232.

67. Biarnès, *Les Français en Afrique Noire*, 368.

68. A. Foures, "Caractères des Interventions Militaires Françaises Outre-Mer de 1960 à nos Jours," *Revue Historique des Armées* 169 (1987), 94.

7

The Intellectual Dimension of Soviet (Russian) Operational Art

David M. Glantz

The Soviets consistently viewed history as a process of dialectical change in nature and society. The discipline of history was a science, which, in their view, "studies the development of human society as a single natural process, regular in all of its great variety and contradictions."[1] This process often produced war, a sociopolitical phenomenon characterized as a continuation of politics by violent means. Anticipating the possibility of war, nations created armed forces to use as "the chief and decisive means for the achievement of political aims, as well as economic, diplomatic, ideological, and other means of struggle."[2]

Given the importance of war, the Soviets approached its study scientifically and systematically within the framework of what they termed "military science," one of many sciences that helped explain the historical process. Soviet military thought developed as much on a theoretical basis as it did from military practice. Over time, Soviet military theorists created concepts and terms associated with a hierarchical, complex range of issues extending from national military policy and military doctrine to finite battlefield tactics. The entire semantic and intellectual hierarchy, beginning with military policy, originated from, reflected, and received official sanction from Communist Party dogma and decision. Although that dogma and the party which propagated it were discredited in the revolution of 1991, the military conceptual hierarchy will likely endure, whether or not the ideology remains discredited.

At the apex of this hierarchy is military policy (*voennaia politika*), the military facet of national policy associated with the use or threatened use by states of the military instrument to achieve national objectives. The use of the armed forces in war and the definition of the nature of war is the purview of military doctrine (*voennaia doktrina*), which, in turn, examines two fundamental components: the political-social and military-technical. Military doctrine, so defined, has combined "scientifically founded views" of military science with official party sanction and, in so doing, unites the objective findings of military analysis with perceived objective truths of socialism.[3] In the broadest sense, even in the absence of a socialist context, future military doctrine will likely reflect those political realities that conditioned the political, economic, and social development of all states of the former Soviet Union.

Within the context of military doctrine, military science (*voennaia nauka*) is "a system of knowledge concerning the nature and laws of war, the preparation of the armed forces and nation for war, and the means of conducting war."[4] Its basic subject is the investigation of armed conflict in war, and, while the state's political leadership manages war, the military leadership and General Staff play a more significant role in the conduct of armed conflict.

Military art (*voennoe iskusstvo*), the main component of military science, is concerned with "the theory and practice of preparing for and conducting military operations on land, at sea, and in the air."[5] The growing complexity of warfare in the twentieth century dictated the necessity for further refinement of terminology describing the levels and scope of military art. This refinement led the Soviets to subdivide military art into the interrelated fields of strategy, operational art, and tactics, each of which described a distinct level of warfare measured against such standards as mission, scale, scope, and duration of military actions. Since "the state of military art depends on the levels of the development of production and means of armed conflict, as well as the nature of social structures," and reflects "the historical and national characteristics of a country, its geographical conditions, and other factors," the definition and relative importance of its subordinate fields of strategy, operational art, and tactics changed over the years since the formation of the Soviet state and will continue to change in the future.[6] A central feature of Soviet and Russian military art is basic, yet evolving, principles governing the nature of armed conflict; these principles have developed in consonance with those influences affecting military art in general.

Since the 1920s, Soviet military theorists have considered military strategy (*voennaia strategiia*) as the highest level of military art, "embracing the theory and practice of preparing the nation and armed forces for war, and planning and conducting strategic operations and war as a whole."[7] Military strategy dominates the other components of the art of war, defining their tasks and the methods of forces on an operational and tactical scale. In turn, military strategy relies upon operational art and tactics, taking into account their capabilities and exploiting their achievements in the performance of strategic (their war-winning) tasks.

The second level of military art is the operational level, identified by the Soviets in the 1920s and used thereafter for the analysis of armed conflict as an intermediate link between tactics and strategy. Operational art (*operativnoe iskusstvo*) encompasses the theory and practice of preparing for and conducting combined and independent operations (*operatsiia*) by large formations of the armed forces. "Stemming from strategic requirements, operational art determines methods of preparing for and conducting operations to achieve strategic goals." In turn, operational art "establishes the tasks and direction for the development of tactics."[8]

Tactics (*taktika*), the lowest level of military art, studies problems relating to battle (*srazhenie*) and combat (*boi*), the basic building blocks of operations. Tactics "investigates the rules, nature, and contents of battle and works out the means of preparing for and conducting battle."[9] Tactics is dialectically interrelated with operational art and military strategy. Strategy determines the nature and methods of conducting war and the place of combat in warfare, while operational art determines the specific tasks that tactics must address. Conversely, tactics influences operational art and military strategy.

This well-articulated system for the study of war emerged in the 1920s and has persisted for the ensuing 70 years. Since the 1920s the fundamental relationship between the levels of war has not changed. However, the scope and importance of each level has varied according to political and military circumstances and, most importantly, to military-technological changes. Moreover, definitions of operational art and retrospective analysis of operational art in the past have been altered to accord with contemporary and anticipated future circumstances. This constant process of analysis and redefinition of the past both reflects and conditions contemporary interpretations of operational art and paves the way for definition of operational art in the future.

It is also important to understand that a gap has always existed in the Red (Soviet) Army between theory and practice. While theorists have routinely propounded some of the most advanced concepts for conducting war, practitioners have blundered on the battlefield, often at catastrophic cost to soldiers and the state alike. In short, the same system that nurtured advanced theory inhibited its artful application in battle with often tragic consequences.

THE FORMATIVE YEARS (1927-1941)

Operational art emerged slowly as a distinct category of military art in the twentieth century. The changing nature of war and its increasing complexity rendered traditional definitions of strategy and tactics less relevant. As understood by nineteenth-century military theorists, war as a series of battles (or large, single engagements) was the object of study for strategy, and battle was the object of study for tactics. Successful battle, which destroyed or incapacitated an enemy's forces, permitted successful achievement of strategic war aims.

Forces unleashed by the political, social, and economic turmoil of the French Revolution and age of Napoleon altered the nature of war. Emerging multiple mass armies, economic mobilization of the state for war, and less limited wartime objectives (often involving the outright destruction of opposing political, economic, and social systems) complicated the traditional framework for analyzing and studying war. Nineteenth and early twentieth century technological innovations facilitated mobilization and employment of ever larger armies and the application of ever increasing amounts of firepower on the battlefield. Combined with a "democratization of war" and the emergence of mass armies, this produced the carnage of mid- and late nineteenth-century and early twentieth-century wars. Nineteenth-century military theorists recognized and wrestled with these changes. Clausewitz voiced such new concepts as "absolute war" and "moral elements of war." Jomini attempted to capture war's increased complexity by describing a new realm of "grand tactics." Military operations matured to a grander scale and took the form of a series of consecutive and mutually related battles fought over a more protracted period of time. Nonetheless, military leaders still planned for and sought to conduct the single battle of annihilation designed to produce decisive strategic results. Single battles of annihilation, however, failed to produce strategic results. The destruction of single armies no longer ensured war termination. While some commanders learned this hard fact in the midst of war, it took the appalling human and materiel losses of the First World War to bring this fact home to most European military theorists. The Soviets claim credit for having been the first nation to recognize the changing nature of war and the first to adjust their military art to meet the new realities: "To its credit, Soviet military theoretical thought, having first succeeded in seeing these tendencies in the development of military affairs, correctly perceived and revealed the new component of military art— operational art."[10]

Operational art, as a distinct field of study, emerged in the 1920s and evolved in the 1920s and 1930s as Soviet military theorists pondered the nature of modern war and solutions to the dilemmas of the First World War, the most important of which was how to restore mobility and maneuver to a stagnant battlefield and to harness those means to achieve strategic aims. Within the framework of major doctrinal and strategic debates, Soviet military theorists, many of them ex-Tsarist officers, tapped their repository of military experiences (the Russo-Japanese War, the First World War, and the Civil War), thoroughly read and studied past and contemporary Western theorists, and shaped a new understanding of the nature of modern war.

The debate over strategy was most fruitful. Spurred on by traditional military thought now tinged with ideological ardor, M. N. Tukhachevsky and others advanced a strategy of annihilation, whereby modern forces equipped with modern weaponry could crush an enemy and quickly achieve strategic ends.[11] Others like A. A. Svechin and N. E. Varfolomeev, cautioned restraint and the adoption of a strategy of attrition to better equip the state (especially a technologically backward one) to survive the appalling destructiveness of modern war.

Svechin, an ex-Tsarist general staff officer and preeminent military thinker, drew heavily on European and Russian military intellectual traditions. His perceptive study of (and his participation in) the Russo-Japanese War, the First World War and the Civil War uniquely equipped him as a premier strategist and virtual creator of the field of operational art. Having joined the Red Army in March 1918, he soon became chief of the All-Russian Main Staff. After the war he joined the faculties of the Frunze and General Staff Academies, where he was professor of staff service, strategy, and military art. His important works included *Strategiia* (Strategy, 1923 and 1927), in which he provided the first and clearest definition of operational art, *Strategiia v trudakh voennykh klassikov* (Strategy in works of military classics, 1927), *Evolutsiia voennogo iskusstva* (The Evolution of military art, 1927-1928), *Klauzevits* (Clausewitz) (1935), and *Strategiia XX veka na pervom etape* (Strategy of the twentieth-century in its first stage, 1937).[12] Svechin became a victim of the purges after 1937. Svechin's collective works—*Strategy*, in particular—provided the basis for, and definition of, operational art as well as an unsurpassed explanation of the context in which operational art was born and would evolve.

Varfolomeev, another an ex-Tsarist officer, served in the Red Army from 1918 as chief of an army staff, deputy front chief of staff, and, later, colleague of Svechin at the Frunze Academy's Department of Strategy. He shared many of Svechin's strategic and operational views and was an active writer of military theoretical books including *Udarnaia armiia* (The shock army) and articles in the military journal *Voina i revoliutsiia* (War and revolution).[13] Varfolomeev focused on German army operations in 1914 and 1918, his work providing the basis for the emerging Soviet concept of successive operations. Although the full measure of this debate is beyond the scope of this chapter, it was within its context that operational art emerged from the pens of Svechin, Varfolomeev, and others from both contending strategic schools.

As a more sophisticated realm, the art embraced new concepts of war at the operational level, which themselves matured throughout the 1930s. The theory of successive operations, a focal point of analysis by both strategic schools in the 1920s, matured in the 1930s into the twin concepts of "deep battle" *(glubokii boi)* and "the deep operation" (*glubokaia operatisiia*), concepts which remained "ideals" of Soviet operational art for 60 years.

The renaissance in Soviet military thought, which gave birth to the field of operational art and the twin concepts of deep battle and the deep operation, and which prompted wholesale Soviet experimentation with new and advanced force structures (for example, motor-mechanized and airborne), continued until 1937. The persistence and originality of these ideas was remarkable given the political repression which swept across the Soviet Union in the 1930s. In 1937, however, the purges struck the military, crushing originality of thought and claiming the lives of many of the Soviet Union's most imaginative military theorists.

Some thinkers survived the purges. But they were few and remained in imminent danger of being swept away by the tide of obsequious kowtowing to Stalin and his victorious cronies. One survivor, G. S. Isserson, a student of operational art and the deep operation, was still alive in the 1970s, when his purged comrades were rehabilitated and their ideas restored to their former state of grace. Isserson was a prestigious theorist and prolific writer, who authored several major books including *Evoliutsiia operativnogo iskusstva* (The evolution of operational art, 1932 and 1937), *Osnovy oboronitel'noi operatsii* (The basis of the defensive operation, 1938) and *Novye formy bor'by* (New forms of struggle, 1940).[14] He was chief of the Operations Department of the Frunze Academy and, later, chief of the Operations Department of the General Staff

Academy. How he was able to write advanced and visionary works as he did and survive in the process is still a mystery. In the 1970s, he wrote several retrospective articles critiquing the work of Soviet military theorists of the 1930s and exposing the dearth of imaginative work done after 1937. In his writings, Isserson explained the essence of the operational level and the requirements for operational success in future war, namely, the capability of conducting deep battle and the deep operation.

The purges accentuated an already existing truth in Soviet (and perhaps Russian) development—the tendency for practice and reality to lag significantly, often disastrously, behind theory. Although operational art emerged as a vibrant new field of military study, most of the operational concepts associated with it were stillborn or only partially developed. The Red Army would discover this truth and suffer mightily as a result of it during the opening months of war in 1941.

The works of Svechin, Varfolomeev, Isserson, and others displayed both the imaginativeness and the futility of Soviet operational theory in the interwar years. Their descendants today still ponder the lessons of what occurs when political folly renders irrelevant imaginative military thought.

THE TEST OF WAR (1941-1945)

While theoretical concepts of operational art dominated the attention of the Soviet military establishment during the interwar years, operational realities and practices plagued Soviet military planners after 1938 and, understandably, preoccupied Soviet military theorists after 22 June 1941. The chief motivating force for the Soviet military was, at first, its defense and, then, its survival and the survival of the state.

Dismal Soviet military performance was clear during the crises and wars which preceded the German invasion of the Soviet Union (notably the Czech crisis, the invasion of eastern Poland, and the Russo-Finnish War); and the catastrophic course of the initial period of war from June until December 1941 confirmed Soviet lack of mastery of the operational realm. Despite sound theoretical concepts, few, if any, Soviet commanders at any level could implement them in the field. The ensuing disasters were strategic in scale and consequence. Retrospective Soviet analysis concluded,

> Commanders and staffs were not fully familiar with all the theories of conducting deep battle, and there were shortcomings in the material base that hindered its realization. Thus, during the war it was necessary to reassess and clarify some aspects of preparing and conducting offensive operations and decide anew many questions on the conduct of defensive operations on a strategic and operational scale.[15]

These questions were addressed anew under the immense pressure of combat conditions and as a part of a quest for survival. The German attack of June 1941 achieved strategic, operational, and tactical surprise and encountered only a partially prepared Soviet strategic defense. Soviet command and control was inept, as Soviet front and army commanders failed to establish coherent defenses and displayed an alarming propensity for launching counterattacks that were predestined to miscarry. Disaster after disaster finally drove the Soviet High Command to seek practical remedies to these problems. The imperative of an ongoing war dictated the need for practical, rather than theoretical, solutions. For four years of war, battlefield practice preceded theory as the Red Army relearned how to operate at the strategic, operational, and tactical levels of war. For this reason, while theoretical writings on operational art dwindled in number, practical works on the conduct of war at all levels blossomed.

In November 1942 the Soviet High Command and General Staff established a mechanism for systematically collecting and processing war experiences, primarily from front and army command levels. This elaborate and effective system ultimately produced hundreds of volumes of secret or top secret analyses of operational techniques and countless other classified series on tactical issues compiled by front commands and force branches.[16] These analyses provided the basis for new regulations, orders, directives, and instructions for the wartime employment of all types of forces. The resulting volumes educated the Red Army in the techniques of modern warfare and made possible the transformation of the Red Army from a force barely able to survive in 1941 into the victorious army of 1945. These documents, and subsequent ones on wartime experience prepared after the war, reflect the practical rebirth of operational art and the fulfillment of those theoretical writings of the 1920s and 1930s. In fact, during the war, the Red Army finally realized its theory of "the deep operation."

Three examples illustrate Soviet approaches to operational art during the war. The first, by Marshal S. K. Timoshenko, expressed the General Staff's hope that reforms carried out in 1940 and 1941 had placed the Red Army on an adequate wartime footing; the second, by Brigade Commander P. D. Korkodinov, illustrated acute Soviet appreciation of what was occurring in 1941; a third, by Major General N. Talensky, surveyed Soviet wartime lessons learned.

Timoshenko, a close associate of Stalin and Voroshilov, was minister of defense in December 1940 when he gave his closing speech to a controversial military conference in Moscow. Although the purges had done their work and discredited the theorists of the 1930s, it is remarkable how much of the intellectual legacy of operational art and the deep operation was evident in his speech. The speech resembles a mini-*ustav* (regulation) on the conduct of operations.[17] By this time, Timoshenko had the task of reforming the Red Army after its poor performance in Poland and the Russo-Finnish War. Despite the remnants of original thought in Timoshenko's speech, Red Army performance in 1941 showed the parlous state into which it had fallen.

Writing in the General Staff journal, *Voennaia mysl'* (Military thought), Korkodinov surveyed the course of the Polish-German war and the beginning of the Second World War in Western Europe. He reached candid and frightening conclusions.[18] His and companion articles typified an assessment that was striking in their acute appreciation of what was occurring in 1940 and 1941.

Tragically, the Red Army did not convert this appreciation into sound military practice.

Talensky, a preeminent Soviet wartime and postwar writer on issues of military strategy and operational art, wrote numerous articles in *Voennaia mysl'* which were noteworthy for their high quality. What is surprising is that he was able to write openly despite the looming presence of Stalin, who, even in wartime, tended to stifle creative thought and claim credit for all military innovation, however slight. Talensky's articles surveyed the state of operational art in 1945, cited those few prewar military theorists who had not been discredited, and resurrected the concept of the deep operation (without resurrecting the memories of its creators). He correctly concluded, "Our operational art has amassed the richest experience, which has permitted in theory and, of necessity, in practice further steps in the development of that most important branch of military art."[19] Talensky's subsequent admonition that the nature of war was everchanging and that further study of wartime experiences was essential to a mastery of operational art in the future set the tone for the subsequent Soviet approach to operational art during the first postwar period.

THE STALINIST POST-WAR YEARS (1946-1953)

In the immediate postwar years, Soviet concern for the operational level of war intensified. Stalinist controls over open and detailed discussion of operational matters in written works produced the outward appearance of atrophy in Soviet military science. Most general texts and shorter articles in open journals deferred to Stalin's role in military science and stressed the universal application of Stalin's permanent operating factors to matters of war. Expressed as lasting principles which determined the course and outcome of war, these included stability of the rear, army morale, the quantity and quality of divisions, the armament of the army, and the organizing ability of command personnel.[20]

The apparent retrenchment in military art was real, a product of native Stalinist suspicion and censorship. While closed source writings do suggest some of this retrenchment, they also reveal the continuing development of military thought despite Stalin's dominance. As recently released archival materials now demonstrate, candid General Staff and General Staff Academy analyses of wartime operations continued unabated.[21] Soviet military theory and operational art also evolved in logical consequence of Great Patriotic War experiences, and the Soviet armed forces were restructured and reequipped in consonance with evolving requirements of operational art and accelerated postwar technological change. There were, of course, certain topics which military theorists were constrained from addressing. These included the politically sensitive issues of surprise, particularly regarding the circumstances of German success in June 1941; the entire topic of the initial period of war; and weaknesses in Soviet strategic and operational defensive theory, which the events of 1941-1942 had made vividly evident. Also proscribed was serious discussion of the impact of atomic weaponry on future warfare, in part because of Stalin's deliberate belittling of the effects of atomic warfare (which, in part, concealed Stalin's real concern for the subject). Aside from these prohibitions, prior to 1953, Soviet theorists could and did address most other facets of operational art, albeit while extolling Stalin's contributions to every positive Red Army wartime achievement. After Stalin's death in 1953, the constraints on writing abated and significant discussion began on hitherto proscribed military issues.

The writings of Lieutenant General V. Zlobin and Major General L. Vetoshnikov in *Voennaia mysl'* typified postwar Soviet attitudes toward operational art.[22] They reviewed Stalin's efforts during the Civil War and credited the Soviet Union with being the first nation to identify the unique operational level of war as opposed to Western experiments with "small strategy" and "grand tactics." The two authors recognized the important theory of deep operations of the 1930s (without mentioning the theorists who developed it) and even alluded to the factor of surprise in June 1941. However, they underscored chauvinistically the superiority of Soviet operational art, which enabled the Soviet Union ultimately to absorb the German blow and to emerge victoriously (due largely, according to the authors, to Stalin's enlightened leadership). The authors' detailed examination of Red Army operational techniques, particularly from 1943 to 1954, demonstrated the postwar dominance of those wartime experiences.[23] Subsequently, Zlobin and Vetoshnikov developed the thrust of their arguments further and placed even greater emphasis on the nature of contemporary and future war, with due deference to Stalin's role as preeminent military theorist and practitioner.

ON THE EVE OF THE REVOLUTION IN MILITARY AFFAIRS

Stalin's death and the subsequent political changes in the Soviet Union—especially the de-Stalinization of the post-1958 period—permitted Soviet military theorists to strip off slowly the veneer of Stalinist principles, which had both insulated operational theory from intensive examination and prevented more active, open discussion of operational questions. It also allowed those theorists to ponder more fully the likelihood and nature of nuclear war. Theoretical debates grew in intensity, paralleled political struggles within the Soviet Union, and culminated in 1960 with full Soviet recognition that a "revolution" had occurred in military affairs. The roots of that recognition were already apparent before 1960.

Writing in 1955, two years after Stalin's death, Lieutenant General A. Tsvetkov traced the evolution of operational art and, in particular, the impact of technological change on its content.[24] More important, he assessed the impact of that change on the operational art of all modern types of combat forces. Refreshingly, references to Stalin's role in the development of operational art virtually disappeared, although broader criticism of the Soviet system remained prohibited. A year later, Colonel V. Vasil'ev surveyed anew the place of operational art in the context of strategy and emphasized the need to develop distinct and detailed theories of operational art relating to the various branches and types of forces.[25] He argued for continued updating of the nature of operational art and, in so doing, reemphasized the factor of surprise, which had been so often ignored in writings during the Stalinist era.

Even more importantly, Marshal of Tank Forces P. A. Rotmistrov, an illustrious former wartime commander and chief of Soviet armored and mechanized forces during the last years of the Great Patriotic War, wrote an article which was the first to recognize the growing influence that atomic weaponry was having on operational art.[26] Rotmistrov's article was transitional, standing astride two periods; one leg was planted in the postwar celebration of wartime operational art while the other probed hesitatingly the nuclear future and what it would mean for the conduct of war. Rotmistrov focused attention on new weaponry of mass destruction, a topic which would preoccupy Soviet theorists for decades to come.[27]

These initial post-Stalin theoretical writings contained many of the same details on the evolution of operational art as had the works of the immediate postwar theorists. What had changed was the incessant reference to Stalin's role in that evolution. Again reflecting ongoing political debates, theorists now placed greater emphasis on the potential impact of technological change on military art and considerably less emphasis on the role and utility of the Stalinist era "permanently operating factors" in war.

THE REVOLUTION IN MILITARY AFFAIRS (1960-1964)

Generally speaking, the revolution in military affairs did not alter appreciably the Soviet definition of operational art. It did, however, diminish its importance in regard to questions of strategy, and, in particular, it lessened concern for conventional operational techniques and increased concern for strategic nuclear concepts. The ensuing period encompassed two distinct stages. The first, lasting from 1960 until roughly the time of Nikita Khrushchev's removal from power in 1964, was marked by intensive de-Stalinization and extensive concern for global nuclear war. Best characterized by Colonel General S. V. Sokolovsky's work *Voennaia strategiia* (Military strategy), Soviet theory during this period discounted the likelihood of conventional war and argued that future war would be inherently and globally nuclear.[28] This belief was underscored by a restructuring of the military to deemphasize operational (ground) forces and instead emphasize nuclear (strategic rocket) forces. This policy was not altogether acceptable to military circles.

Writing in 1961, Major General B. Golovchiner typified Soviet attitudes toward operations in a nuclear context.[29] Stressing the increased importance of joint operations, he emphasized the emerging role of atomic weapons, rocket delivery systems, and radio-electronics in modern combat. These new systems, in turn, placed an even greater premium on depth of operations and the concept of simultaneous engagement of enemy forces. The following year Colonel I. Marievsky traced in detail the pre-1920s roots of operational art and partially rehabilitated the concepts and reputation of Svechin, the long-ignored father of operational art.[30] Marievsky provided a wealth of material on developments during the interwar years hitherto unavailable in print and, in particular, candidly addressed the damage done to Soviet military thought by Stalin's excesses.

In 1963, to demonstrate further the process of de-Stalinization and the new period of *glasnost'* under Khrushchev, *Voenno-istoricheskii zhurnal* (*VIZh*) (Military-historical journal) republished a 1932 exposition by Red Army Chief of Staff A. I. Egorov on operational art and tactics.[31] Publication of the article by Egorov, a victim of the purges, marked a new commitment to reinvestigate the impact of the purges on operational thought and assess the role that repression had on the disastrous initial stages of the Great Patriotic War. Set within the context of Soviet fixation on nuclear matters, this new preoccupation with the failures of the late 1930s had the added effect of revitalizing Soviet concern for operational art. The subsequent removal of Khrushchev, in part prompted by the dissatisfaction of senior Soviet officers over the reduction of the ground forces' influence (and the concomitant deemphasis of operational matters), set the stage for a fundamental reassessment of military doctrine and renewed concern for conventional war and operational art.

REINVIGORATION OF OPERATIONAL ART (1965-1970)

Articles and studies published during the period after 1964 evidence a movement away from the preoccupation with nuclear questions: first, they examine the historical roots of operational art; and, second, by 1968, they demonstrate acute awareness that traditional operational techniques applied within a modern context. These articles were, in fact, precursors to the period of the 1970s, when operational art and conventional operations in a "nuclear scared" context again became preeminent Soviet military concerns.

A capstone Soviet work on the development of operational art appeared in 1965. Entitled *Problemy strategii i voennogo iskusstva v sovetskikh voennykh trudakh (1917-1940)* (Problems of strategy and operational art in Soviet military works), the two-volume work contained selections from the writings of a host of interwar theorists, many of which had not been available to readers since the 1930s.[32] The introduction to the book by the chief of the Soviet Army General Staff, Marshal of the Soviet Union M. V. Zakharov, emphasized the importance of the writings and officially sanctioned the rehabilitation process of such writers as Svechin, Tukhachevsky, and a host of forgotten or scarcely remembered theorists on operational art.

Also during 1965, *VIZh* published a lengthy retrospective account of the development of operational art during the 1930s by the surviving theorist of the interwar years, Isserson.[33] The article provided fresh details on the process by which operational art was developed, exposed the views and contributions of those who were purged in the late 1930s, and candidly explained for the first time how Stalinism had adversely affected the process. A subsequent article written by Colonel A. Golubev complemented and critiqued the piece by

Isserson and also fully rehabilitated Svechin, the key strategic and operational theorist.[34] Prompted by the ongoing debate over the nature of both nuclear and conventional war, Golubev provided some historical context by resurrecting the parameters of the great strategic debate of the 1920s regarding the contending strategic schools of "attrition" and "annihilation." His writings thoroughly examined the works of Tukhachevsky, Triandafillov, and others, and cast new light on the concept of successive operations.

This debate continued to mature when, in 1966, Major General N. Pavlenko published a substantial piece in *VIZh*, which addressed the strategic context for the development of operational art in the 1920s and again reviewed in detail the writings and views of purged military theorists, in particular Svechin and Tukhachevsky.[35] Pavlenko's article focused on the evolution of strategic offensive concepts within the parameters of the strategic debate and highlighted the role of successive operations. The Chief of Soviet Ground Forces and Deputy Minister of Defense Army General I. Pavlovsky subsequently applied the same critical eye to the development of operational art in the Great Patriotic War. Pavlovsky's article reflected a still broader trend as major new books appeared on the subject, for instance, Colonel General P. A. Kurochkin's, *Obshchevoiskovaia armiia v nastuplenii* (The combined-arms army on the offensive).[36]

Completing the reinvigoration of the analytical basis of operational art, Marshal of the Soviet Union M. V. Zakharov, chief of the Soviet General Staff, published an article in 1970 that reviewed in detail the historical development of the concept of the deep operation, the core element of historical operational maneuver.[37] Zakharov underscored the contemporary utility of the deep operation and paved the way for the Soviet fixation on operational maneuver, that would dominate operational theory for the next decade and a half. The intellectual ferment is clear in these articles, as is the steady movement away from an obsession with nuclear warfare to an increased faith in the utility of conventional operations, a faith that undergirded renewed Soviet interest in operational art.

THE HEYDAY OF OPERATIONAL ART (1970-1986)

Soviet fixation on nuclear war and strategy, and the resulting eclipse of operational art, had entirely eroded by 1970. Heightened concern for operational art in the 1970s, accompanied by Soviet efforts to restructure their armed forces in order to improve their operational capabilities, elevated the importance of that field from its relative position of neglect in the early 1960s to a major area of concern. While Soviet military theorists agreed that the introduction of nuclear weapons had significantly altered the nature of future war and the contents of operational art (and operations), they renewed their faith in the validity of operational art as a key subject in the mastery of warfare. Consequently, they reinvestigated the key subject of the initial period of war, redefined traditional aspects of mass and concentration, and focused on the conduct of maneuver (both operational and tactical) designed to lessen the likelihood that nuclear weapons would be used in future war and, if they were used, to lessen the effects of these weapons (particularly tactical nuclear weapons). Throughout the 1970s, Soviet study of maneuver focused on antinuclear (*protivoiadernyi*) maneuver and culminated in development of the twin concepts of the theater-strategic offensive and operational maneuver by operational maneuver groups (OMGs).[38] Against this backdrop, Soviet writings on all aspects of operational art and operational maneuver broadened and intensified.

Writing in 1970, Major General M. Cherednichenko identified contemporary characteristics of operational art, including a careful distinction between nuclear and conventional war, the issue of local war, the concept of theater-strategic offensives, and new technological requirements such as the mathematical solution of problems of military art.[39] Soon thereafter, Lieutenant General I. Zav'ialov and Colonel V. Chervonobab sketched out the context, parameters, and nature of operational art and set the tone for subsequent writings in the 1970s. The former returned to the nuclear theme, analyzed the effects of nuclear warfare on traditional relationships between strategy, operational art, and tactics, and outlined the possible impact of nuclear war on specific aspects of operational art.[40] The latter assessed the impact of potential nuclear war on the laws of war and principles of military art.[41]

The first volume of the new *Sovetskaia voennaia entsiklopediia* (Soviet military encyclopedia), appearing in 1976, included a substantial article written by Colonel General N. V. Ogarkov on the nature and significance of the deep operation.[42] The fact that Ogarkov was then First Deputy Minister of Defense underscored Soviet concern for the subject. Ogarkov's writings, and a host of articles on operational art and operational maneuver which followed, reflected development within the General Staff of the OMG concept. In 1978 Colonel L. I. Voloshin returned to the theme of the deep operation, relating its development and applying it to contemporary and future operations. He concluded,

> Although the term "deep operations (battle)" has not been used in official documents since the 1960s, the overall principles of that theory have not lost their meaning, and on the contemporary materiel base of armed struggle they continue to perfect themselves.[43]

Without directly saying so, he provided necessary context for the development and employment of modern operational maneuver groups.

That same year N. N. Fomin reevaluated the periodization of operational art, underscoring the increasing scope and complexity of modern offensive operations.[44] Subsequent articles intensified the study of operational art and operational maneuver by analyzing new aspects of operational art. Major General V. F. Mozolev and Colonel General M. I. Bezkhrebty focused on the combined operation and its role on the complex modern battlefield, emphasizing the necessity for better command and control, use of mathematical modeling, and the growing significance of long-range fires and deep rapid maneuver.[45] Meanwhile, Colonel R. Savushkin provided a more historical context by surveying the evolution of the term "operation" during the pre-1930s period and appealing for more intense future study of the topic.[46]

While Soviet preoccupation with operational art and the combat utility of operational maneuver continued into the 1980s, technological developments and the manner in which foreign armies exploited them generated renewed concern and reaction from Soviet military theorists in both the operational and tactical realms. In the early 1980s, technological realities forced Soviet military theorists to address a whole range of new combat problems associated with evolving combat weaponry. The first problem was the appearance on the battlefield of longer-range, high-precision weapons (*vysokotochnoe oruzhie*), more lethal and sophisticated descendants of the anti-tank guided missiles (ATGMs) of the 1970s. Simultaneously, Western acceptance of the operational level of war as a valid concept and the development of new maneuver concepts compounded the adverse effects of this technological revolution on traditional Soviet offensive concepts. The U.S. concept of AirLand Battle and NATO's concept of follow-on-forces attack (FOFA) sought to capitalize on the new weaponry by conducting "deep battle" to strike enemy forces to the depths of their formation. These essentially operational concepts placed Soviet second echelons, OMGs, and rear area facilities in increased jeopardy during future war. In short, the new weaponry and Western operational concepts forced the Soviets to abandon, or at least seriously alter, traditional operational concepts for echelonment and maneuver of forces.

In the late 1970s and early 1980s, threatened by possible enemy wartime use of tactical and theater nuclear weapons, Soviet military theorists recommended wartime use of shallower strategic and operational echelonment: in essence, a single echelon of fronts, each with the preponderance of its armies also formed in single echelon. They believed that wartime employment of OMGs would solve the problem by replacing cumbersome second echelons with more dynamic, flexible, and rapidly maneuvering exploitation forces that could cope with the nonlinear nature of combat at the operational level of war.

The Soviets ultimately countered Western introduction of new high-precision weaponry in the early 1980s by almost totally abandoning linear concepts of warfare, even at the tactical level. By 1984, although Soviet theorists had defined the problems facing future operational art and tactics, they had not yet found a complete solution. Their attempts to do so beginning in 1985 would usher in a more complex period of military development—a period made even more difficult by ensuing political, economic, and social problems that shook the foundation of the Soviet empire and state and compounded the difficulties already encountered in mastering the subject of future war.

The development and fielding by Western armies of high-precision weaponry, more sophisticated cybernetic systems for command and control, and information processing and dissemination posed an immense challenge to Soviet theorists, whose own scientific establishment simply could not compete with their Western counterparts. What was clear was that basic operational and tactical techniques, as well as force structures, would have to change to meet the new demands of what was increasingly becoming a fragmented, nonlinear battlefield.

Several trends were notable as theorists struggled with these new problems. First and foremost, they placed even heavier emphasis than before on maneuver at both the operational and tactical levels. Second, traditional concepts of concentration and echelonment had to evolve to meet new requirements. Much of this discussion took place within the well-established framework of debate over the historical evolution of operational art. In the early 1980s, Soviet theorists intensified their study of successive offensive operations as the key to combat success in a prospective theater-strategic operation. Colonel R. A. Savushkin surveyed the root of successive offensive operations during the interwar years, while Colonel General M. I. Bezkhrebty did likewise for the post-1941 period.[47] These articles on the nature of the deep operation laid the groundwork for a more productive debate of contemporary issues. An article by Colonel P. G. Skachko in 1985 typified that extensive debate. He examined modern requirements for deep operations, incorporating into his analysis important new means of long-range fire (high-precision weapons) and emphasized the growing role of air-mobility. Soon Soviet military theorists would argue that the increased importance of air-mobility had made combat three-dimensional and required the creation of an air echelon capable of conducting "land-air" battle in concert with ground echelons of operational and tactical maneuver forces.[48]

THE DEBATE OVER DEFENSIVENESS (1987-1991)

By the end of 1985, which Soviet tactical specialist Lieutenant General V. K. Reznichenko identified as the end of an old and the beginning of a new period of military development, Soviet military theorists were facing military dilemmas exacerbated by growing economic problems.[49] In reality, most of these issues related to the overwhelming costs associated with the continued militarization of the state and economy. The prospective quickening pace of combat resulting from improved force mobility and the burgeoning lethality and accuracy of weaponry called into question long-held assumptions regarding the nature of future ground combat. The Soviets still adhered to the general concept of the theater-strategic operation, and Soviet theoretical writings evidenced an abiding faith in the offensive as the best guarantee of victory in future war. Central to that concept were the traditional aspects of deep battle, deep operations, and the vehicle for operational maneuver, the OMG.

Major problems, however, plagued Soviet theorists. The first was that of adjusting operational concepts to address the battlefield presence of high-precision weaponry. The second was to counter Western concepts of deep battle. The initial Soviet solution to both problems was their near total abandonment of linear concepts of warfare. Soviet theorists advanced new concepts of nonlinear war, identifiable down to the lowest tactical level; these were characterized by the adoption of new echelonment techniques, the formation and employment of tailored combined arms forces, increased frequency of independent actions by operational and tactical forces, and a proliferation of air assault forces (an air echelon) at every level of command. As recently as 1987, the concept of antinuclear operational maneuver still provided a cornerstone for Soviet operational and tactical techniques designed to preempt, preclude, or inhibit the enemy from resorting to nuclear warfare. Concurrently, Soviet analysts had concluded that high-precision weapons essentially posed the same threat to attacking forces as had tactical nuclear weapons, and that even greater emphasis on operational and tactical maneuver would be a partial remedy to countering enemy use of high-precision weaponry. To capitalize fully on the effects of maneuver, the Soviets believed that they had to reduce planning time and execute command and control more precisely. This required increased emphasis on the use of cybernetic tools, including automation of command and expanded reliance on tactical and operational calculations.

The Soviet concept of nonlinear *(ochagovyi)* combat was the nucleus of a larger Soviet concept of "land-air battle," which had evolved by 1988 in juxtaposition to the early U.S. concept of AirLand Battle. Nonlinear war in no way conflicted with traditional Soviet operational concepts like deep operations and represented but another stage in a long evolution from the 1930s. A 1988 article by Colonel V. I. Ulianov exemplified continued Soviet analysis of deep operations and reaffirmed its contemporary applicability.[50]

Sharp changes, however, soon occurred, halting this evolution in its tracks. These changes were prompted not by military necessity but, rather, by the worsening Soviet economic and political situation. Military-theoretical writings remained remarkably evolutionary and emphasized the offensive right until 1987, when the Soviet political and military leadership announced a fundamental shift to a defensive military doctrine. Understandably, the General Staff journal, *Voennaia mysl'*, first reflected altered military doctrine. Whereas, prior to 1987, this journal had consistently published two to four times as many articles on offensive themes as on defensive ones, in 1987 the ratio began shifting in the other direction. By 1990, defensive articles outnumbered offensive ones by a ratio of three to one. Other military journals followed suit.

In 1987 Colonel R. A. Savushkin reexamined Soviet defensive thought in the interwar years, a trend which culminated the following year with the full rehabilitation of Svechin, the partially rehabilitated defensive specialist of the 1920s.[51] Indeed, the lead article in the January 1988 issue, which appeared as an unsigned editorial, provided the rationale for defensive doctrine and explained its implications for military art.[52] A subsequent article by Colonel E. G. Korotchenko carefully knit the principles of military and operational art into the fabric of defensiveness.[53]

Many military theorists were less than enthusiastic about this trend as they attempted to accommodate older offensive themes with the requirements of their new defensive doctrine. Writing in the December 1988 issue of *Voennaia mysl'*,

Army General G. I. Salmanov gently reminded readers that operational art and the nature of modern war did not change overnight and that defensive doctrine was essentially a new political approach to the problem of war and global stability.[54] Salmanov's warning was vividly underscored in a 1991 article by Colonel General I. N. Rodionov, the commandant of the Voroshilov General Staff Academy, which, with other important writings, represented a reserved approach to the wisdom and feasibility of too much defensiveness.[55] At the same time, the debate continued as the preeminent Soviet tactician, Lieutenant General V. G. Reznichenko, presented readers with a balanced assessment of both offensive and defensive army operations, and Colonel A. N. Zakharov wrote a net assessment of the impact on recent technological changes on the nature of armed combat.[56]

Rodionov's article typified the views of many military analysts who questioned the wisdom and validity of too great a dependence on defensiveness; he expressed concern that theorists and planners err on the side of prudence when attempting to define what future force levels and operational techniques suited the needs of defensive sufficiency, first in a Soviet, and then in a Russian context. Since the fall of the Soviet Union, these theorists have tended to adhere to older "Soviet" views on the nature of war—in particular, at the operational level—and, more generally and ominously, the role and function of the military in the new Russian state.

CONCLUSIONS

Since its creation as a distinct realm of study in the 1920s, Soviet operational art has changed little. In theory and practice, identification of the operational level, its use in planning and conducting war, and its utility for studying the nature of war, retrospectively or as a forecasting vehicle, has proven its worth. In this sense, Soviet operational theorists have contributed not only to their own military development but to the health of military establishments of other nations who appreciated and adopted the Soviet approach. It is no coincidence that Western study of the operational level as a distinct and valid subject in its own right burgeoned in the late 1970s and 1980s with positive results.

The contents, scope, and importance of operational art have evolved over time in consonance with the changing nature of war. Most important, study of the operational level has fostered better understanding of the impact of technological change on warfare at all levels. At times technological changes have increased the importance of operational art (as in the 1930s and 1970s), and at other times major technological innovations (atomic, nuclear) have tended to lessen the importance of operational art relative to strategy. Today, with a new technological revolution in weaponry (that is, high-precision weapons and weapons based on new physical principles), it is important again to anticipate what impact they will have on operational art.

Likewise, technological changes have altered the relative balance and importance of the offense and defense at the operational level. The tank and airplane of the 1930s unfettered the offense and made blitzkrieg and deep operations supreme. Corresponding development of anti-tank defenses during the Second World War restored the viability of the defense until new combined arms concepts empowered the offense with new strength and vigor by 1945. In much the same way, ATGMs of the 1970s seemed to reinvigorate the defense, while operational and tactical maneuver concepts seemed to restore the power of the offense. More recently, the potential effect of high-precision weapons on combat again cast doubt on the viability of tank-based offensive concepts. Finally, political and economic conditions have affected operational art. They did so during the late 1930s, when Stalin's purges stifled creativity in both the theoretical and practical realms, and, again, in the mid-1950s, when Stalin's demise dovetailed with Soviet recognition of the importance of nuclear weaponry. More recently, in the 1980s systemic political and economic atrophy generated a similar atrophy in the military theoretical realm as the Soviet system blundered toward the abyss and ultimate collapse.

These incessant dialectical changes have accorded the operational realm a dynamic and ever changing character and have impelled constant study by military establishments hoping to master the complexities of operational art. The historical development of operational art eloquently attests to the necessity for constant and imaginative study if military establishments are to adjust, survive, and master future war. Today, Russian operational art is being influenced by a number of critical factors. Among the most important are Russian perceptions of the interface of technology and the nature of future war; of particular importance is the influence of high technology-weaponry, which may have rendered some aspects of traditional operational art obsolete. Some Russian military theorists now question the possibility and utility of operational maneuver on a battlefield dominated by this new weaponry. The course and outcome of the Gulf War and more recent cases of low-intensity conflicts have only reinforced these concerns and underscored the inability of the Russian military-industrial complex to compete technologically with its Western counterparts.

Those debating the validity and applicability of operational art generally divide into two schools. The traditional school, represented by such theorists as General of the Army (retired) M. Gareev, one of the creators of the OMG concept, insists that operational art retains its currency and that operational maneuver is feasible in virtually every combat context. Those who challenge the traditional view, such as Colonel General V. N. Lobov, former commandant of the Frunze Academy, do so for a combination of military and political reasons. Some seriously question the feasibility of operational maneuver, and the rationale of an intermediate level of war, when modern weapons produce such rapid resolution of strategic missions. Still others, who can be termed "rejectionists," question the validity of operational art now and in the past simply because of the concept's close association with the hated former Soviet regime.

This ongoing debate in the operational realm, however, reflects even greater uncertainties associated with the future form, and even existence, of the Russian state. As products of the Soviet system, Russian military theorists and their political masters cannot develop military doctrine or define military science or military art and its component elements (strategy, operational art, and tactics) without a clear geopolitical context: that is, they require a clear understanding of the nature, configuration, and international aims of the Russian state relative to its neighbors. Today that context is lacking. Thus, while attempting to come to grips with the nature of future war, Russian military theorists are also wrestling with their own identity and with a host of associated vital military subjects, including threats, military doctrine, military science, military art, strategy, operational art, tactics, and force structure. Unless and until the context is defined, these issues will remain unresolved. In essence, therefore, the fate and survival of both the Russian state and operational art as it is now known lies in the balance.

NOTES

1. "Istoriia" (History), *Sovetskaia istoricheskaia entsiklopediia* (Soviet historical encyclopedia), vol. 6 (Moscow, 1965), 578-90.

2. D. A. Volkagonov and S. A. Tiushkovich, "Voina" (War), *Sovetskaia voennaia entsiklopediia* (Soviet military encyclopedia), 2 (1976), 301, (hereafter *SVE*).

3. *Slovar' osnovnykh voennykh terminov* (Dictionary of basic military terms) (Moscow, 1965) (translated and published by U.S. Air Force, 1977), 37. See also "Doktrina voennaia" (Military doctrine), *SVE* 3 (1976), 225-29.

4. A. A. Grechko, "Voennaia nauka" (Military science), *SVE* 2 (1976), 183-84.

5. S. P. Ivanov and A. I. Evseev, "Voennoe iskusstvo" (Military art), *SVE* 2 (1976), 211.

6. "Voennoe iskusstvo" (Military art), *Voennyi entsikopedicheskii slovar'* (Military encyclopedic dictionary) (Moscow, 1983), 140-41.

7. N. V. Ogarkov, "Voennaia strategiia" (Military strategy), *SVE* 7 (1979), 555.

8. V. G. Kulakov, "Operativnoe iskusstvo" (Operational art), *SVE* 6 (1978), 53.

9. I. G. Borets, "Taktika" (Tactics), *SVE* 7 (1979), 628-34.

10. Kulakov, "Operativnoe iskusstvo, 55.

11. Among Tukhachevsky's many works, see "K voprosu o sovremennoi strategii" (Toward the question of modern strategy), *Voina i voennoe iskusstvo v svete istoricheskogo materializma* (War and military art in the light of historical materialism) (Moscow, 1927), 127-33; and "Novye voprosy voiny" (New questions of war), *Izbrannye proizvedeniia (*Collected works), Vol. 2 (Moscow, 1964), 184-87.

12. Among Svechin's many publications, see *Strategiia* (Strategy), 1st edition. (Moscow, 1926); his edited *Strategiia v trudakh voennykh klassikov* (Strategy in the works of the military classics), 2 vol. (Moscow, 1927); *Evolutsiia voennogo iskusstva* (The evolution of military art), 2 vol. (Moscow, 1927-1928).

13. N. Varfolomeev, *Udarnaia armiia* (The shock army) (Moscow, 1933); and "Strategiia v akademicheskoi postanovke," [Strategy in an academic setting], *Voina i revoliutsiia* (War and revolution), no. 11 (November 1928), 78-93.

14. See G. S. Isserson's *Evoliutsiia operativnogo iskusstva* (The evolution of operational art), 2nd edition (Moscow, 1937); *Osnovy oboronitel'noi operatsii* (The basis of the defensive operation) (Moscow, 1938); and *Novye formy bor'by* (New forms of struggle) (Moscow, 1940).

15. Kulakov, "Operativnoe iskusstvo," 55.

16. For example, *Sbornik materialov po izucheniiu opyta voiny, No. 1-26* (Collection of materials for the study of war experiences, 26 vols.) (Moscow, 1942-1948). Prepared by the Military-Scientific Directorate of the General Staff and classified "secret" or "top secret," these volumes were declassified in 1964 and released to the public in 1989. A 23-volume set of tactical analyses accompanied this operational set.

17. S. K. Timoshenko, *Zakliuchitel'naia rech' narodnogo komissara oborony Soiuza SSR Geroia i Marshala Sovetskogo Soiuza S. K. Timoshenko na voennom soveshannii 31 dekabriia 1940 g.* (Concluding speech of Peoples' Commissar of Defense of the USSR, Hero and Marshal of the Soviet Union S. K. Timoshenko at a 31 December 1940 military conference) (Moscow, 1941).

18. P. D. Korkodinov, "Kharakter sovremennykh boev" (The nature of modern battle), *Voennaia mysl'* (Military thought), hereafter *VM,* no. 2 (February 1941), 72-86. Korkodinov and others wrote sound analyses of German operations in Poland, Norway, and France in both *Military Thought* and *Voenno-istoricheskii zhurnal* (Military-historical journal, hereafter *VIZh*).

19. N. Talensky, "Razvitie operativnogo iskusstva po opytu poslednikh voin" (The development of operational art based on the experience of recent wars), *VM,* nos. 6-7 (June-July 1945), 3-15.

20. For a clear statement of Stalin's "Permanent Operating Factors", see I. V. Maryganov, *Peredovoi kharakter sovetskoi voennoi nauki* (The advanced nature of Soviet military science) (Moscow, 1953).

21. See, for example, the series *Sbornik voenno-istoricheskikh materialov Velikoi Otechestvennoi voiny, vypusk 1-19* (Collection of military-historical materials of the Great Patriotic War, issues 1-19) (Moscow, 1949-1968), which contain processed analytical studies of a variety of Soviet wartime operations.

22. V. Zhlobin, L. Vetoshnikov, "Ob operativnom iskusstve Sovetskoi Armii" (Concerning the operational art of the Soviet Army), *VM,* no. 3 (March 1947), 3-15, no. 4 (April 1947), 3-18.

23. See L. Vetoshnikov, "Operativnoe iskusstva i ego mesto v sovetskom voennom iskusstve" (Operational art and its place in Soviet military art), *VM,* no. 11 (November 1949), 3-12; and V. Zhlobin, "Tvorchestvo sovetskogo operativnogo iskusstva v Velikoi Otechestvennoi voine" (The creative work of Soviet operational art in the Great Patriotic War), *VM,* No. 5 (May 1950), 15-30.

24. A. Tsvetkov, "Operatsiia, ee sushchnosti i znachenie v sovremennoi vooruzhennoi bor'be" (The operation, its essence and meaning in modern armed struggle), *VM,* no. 3 (March 1955), 39-52.

25. V. Vasil'ev, "Operativnoe iskusstvo kak sostavnaia chast' sovetskogo voennogo iskusstva" (Operational art as an integral part of Soviet military art), *VM,* no. 6 (June 1965), 3-13.

26. P. Rotmistrov, "O sovremennom sovetskom voennom iskusstve i ego kharakternykh chertakh" (Concerning contemporary Soviet military art and its characteristic features), *VM,* no. 2 (February 1958), 82-95.

27. P. Rotmistrov, "O roli vnezapnosti v sovremennoi voine" (Concerning the role of surprise in modern war), *VM*, no. 2 (February 1955), 14-26.

28. S. V. Sokolovsky, *Voennaia strategiia* (Military strategy) (Moscow, 1968).

29. B. Golovchiner, "Nekotorye voprosy sovremennogo operativnogo iskusstva" (Some questions of modern operational art), *VM*, no. 10 (October 1961), 41-49.

30. I. Marievsky, "Stanovlenie i razvitie operativnogo iskusstva" (The formation and development of operational art), *VIZh,* no. 3 (March 1962), 26-40.

31. A. I. Egorov, "Taktika i operativnoe iskusstvo RKKA na novom etape" (Tactics and operational art in a new stage), *VIZh*, no. 10 (October 1963), 30-39.

32. *Problemy strategii i voennogo iskusstva v sovetskykh voennykh trudakh (1917-1940)* (Problems of strategy and operational art in Soviet military works 1917-1940) (Moscow, 1965).

33. G. Isserson, "Razvitie teorii sovetskogo operativnogo iskusstva v 30-e gody" (The development of Soviet operational art in the 1930s), *VIZh*, no. 1 (January 1965), 36-46.

34. A. Golubev, "Obrashchena li byla v proshloe nasha voennaia teoriia v 20-e gody?" (Did our military theory of the 1920s pay attention to the past?), *VIZh*, no.10 (October 1965), 35-47.

35. N. Pavlenko, "Nekotorye voprosy razvitiia teorii strategii v 20-kh godakh" (Some questions concerning the development of a theory of strategy during the 1920s), *VIZh*, no. 5 (May 1966), 10-26.

36. I. Pavlovsky, "Sovetskoe operativnoe iskusstvo v Velikoi Otechestvennoi voine" (Soviet operational art in the Great Patriotic War), *VM*, no. 3 (March 1968), 18-32; P. A. Kurochkin, *Obshchevoiskovaia armiia v nastuplenii* [The combined arms army on the offensive] (Moscow, 1966).

37. M. Zakharov, "O teorii glubokoi operatsii" (Concerning the theory of the deep operation), *VIZh*, no. 10 (October 1970), 10-20.

38. See F. Sverdlov, "K voprosy o manevr v boiu" (Concerning the question of maneuver in battle), *Voennyi vestnik* [Military herald], no. 8 (August 1972), 31. A full explanation of the evolution of the notion of antinuclear maneuver into the operational maneuver group (OMG) concept is found in David M. Glantz, *The Soviet Conduct of Tactical Maneuver* (London, 1991), 213-27.

39. M. Cherednichenko, "Nekotorye cherty sovremennogo voennogo iskusstva" (Some characteristic features of modern military art), *VM*, no. 2 (February 1970), 40-50.

40. I. Zav'ialov, "Evoliutsiia c sootnoshenii strategii, operativnogo iskusstva i taktiki" (Evolution in the correlation of strategy, operational art, and tactics), *VM*, no. 11 (November 1971), 25-33.

41. V. Chervonabab, "Printsipy voennogo iskusstva i ikh razvitie" (Principles of military art and their development), *VM*, no. 11 (November 1971), 34-44.

42. N. V. Ogarkov, "Glubokaia operatsiia (boi)" (The deep operation {battle}), *Sovetskaia voennaia entsiklopediia, T.2* (Soviet military encyclopedia), Vol. 21 (Moscow, 1976), 574-78.

43. L. I. Voloshin, "Teoriia glubokoi operatsii i tendentsii ee razvitiia" (The theory of deep operations and tendencies in its development), *VM*, no. 8 (August 1978), 14-26.

44. N. N. Fomin, "Sovetskoe operativnoe iskusstvo: zarozhdenie i osnovnye etapy razvitiia" (Soviet operational art: its birth and the bases of its stages of development), *VM*, no. 12 (December 1978), 16-25.

45. V. F. Mozolev, "Ob obshchikh osnovakh teorii operativnogo iskusstva" (About the overall bases of the theory of operational art), *VM*, no. 3 (March 1979), 13-22; and M. I. Bezkhrebty, "Sovmestnaia operatsiia—glavnaia forma sovremennykh boevykh deistvii" (The joint operation—the main form of modern military actions), *VM*, no. 7 (July 1979), 27-34.

46. R. Savushkin, "K voprosu o vozniknovenii i razvitii operatsii" (Concerning the question of the origin and development of the operation), *VIZh*, no. 5 (May 1979), 78-82.

47. M. I. Bezkhrebty, "Podgotovka posleduiushchikh nastupatel'nykh operatsii" (The preparation of successive offensive operations), *VM*, no. 7 (July 1982), 28-38; and R. Savushkin, "K voprosy o zarozhdenii teorii posledovatel'nykh nastupatel'nykh operatsii" (On the question of the origin of the theory of successive offensive operations), *VIZh*, no. 5 (May 1983), 12-20.

48. P. G. Skachko, "Odnovremennoe vozdeistvie na vsu glubinu operativnogo postroeniia protivnika—vedushchaia tendentsiia v razvitii teorii operativnogo iskusstva" (Simultaneous action against the entire depth of the enemy's operational structure—leading trends in the development of the theory of operational art), *VM*, no. 7 (July 1985), 18-24.

49. V. Reznichenko, "Sovetskie vooruzhennye sily v poslevoennyi period" (The Soviet armed forces in the postwar period), *Kommunist vooruzhennykh sil* (Communists of the armed forces), no. 1 (January 1988), 68-88.

50. V. I. Ulianov, "Razvitie teorii glubokogo nastupatel'nogo boia v predvoennye gody" (The development of the theory of deep offensive combat during the prewar period), *VIZh*, no. 3 (March 1988), 26-33.

51. R. A. Savushkin, "Evoliutsiia vzgliadov na oboronu v mezhvoennye gody" (The evolution of views on the defense during the interwar years), *VM*, no. 1 (January 1987), 37-42.

52. "Oboronitel'nyi kharakter sovetskoi voennoi doktiny i podgotovka voisk (sil)" (The defensive nature of Soviet military doctrine and training of troops {forces}), *VM*, no. 1 (January 1988), 3-13.

53. E. Korotchenko, "Ob evoliutsii printsipov voennogo iskusstva" (On the evolution of the principles of military art), *VM*, no. 9 (September 1988), 22-30.

54. G. I. Salmanov, "Sovetskaia voennaia doktrina i nekotorye vzgliady na kharakter voiny v zashchita sotzializma" (Soviet military doctrine and some views on the nature of war in defense of socialism), *VM*, no. 12 (December 1988), 3-13.

55. I. N. Rodionov, "O nekotorykh polozheniakh sovetskoi voennoi doktriny" (On several tenets of Soviet military doctrine), *VM*, no. 3 (March 1991), 2-9.

56. V. G. Reznichenko, "Podgotovka i provedenie armeiskikh operatsii" (Preparation and conduct of army operations), *VM*, no. 1 (January 1991), 19-27; and A. N. Zakharov, "Tendentsii razvitiia vooruzhennoi bor'by" (Developmental trends of armed struggle), *VM*, nos. 11-12 (November-December 1991), 9-15.

8

Filling the Void: The Operational Art and the U.S. Army[1]

Richard M. Swain

Ideas are important. Born or adopted in particular historical circumstances, they affect man's understanding of his world and, therefore, influence behavior. Ideas are joined to form concepts and concepts are merged to form systems intended to achieve particular purposes. In military organizations, when such systems become unified institutional theories of war fighting, they are called *doctrine*. If the doctrine appears to be effective in interpreting events and guiding action, it is retained and grows. If not, the concepts and ideas are discarded or changed, sometimes in time to avoid disaster, sometimes only after.

An army's doctrine is also a body of ideas and concepts designed to anticipate circumstances a military organization might encounter, and to limit responses to those deemed best most of the time. Doctrine, then, is a means of achieving tactical and operational discipline. It may also be a way to explain requirements to those who organize, train, and equip military organizations. Finally, it is often a means to avoid the errors of the last war in the next. This broader view of doctrine became prominent in the U.S. Army only after the war in Vietnam, with the creation of the army's Training and Doctrine Command (TRADOC) in 1973.

With the activation of TRADOC under General William E. DePuy, doctrine achieved a new importance in the American army. The army underwent a period of intellectual growth between 1974 and 1986. The articulation of the official view of what doctrine should be was accompanied by a public debate of unusual intensity and controversy which found particular voice among European NATO allies. These controversies reflected changing European perspectives of U.S. military intentions as the "malaise" of the Carter administration was succeeded by Reagan bellicosity.

As always, individual men mattered, men both of ideas and of authority. However, it was not enough simply to have a good idea. What was essential was the ability to convert that idea into action, though the sometimes heated debate clearly created an environment in which frequent revision of the institution's basic body of formal beliefs appeared necessary.

Early fascination with operational research modeling as a basis for doctrinal prediction was succeeded by a conservative reaction, a turn to more traditional, theoretical, and historical sources of inspiration, and, ultimately, to a collective wisdom achieved by field trials combined with a few central concepts cribbed from Clausewitz. If the central frustration of the collective memory of Vietnam was recollection of a conflict in which all the battles were won and the war lost,[2] the collective answer seemed to be found with the adoption of the concept of operational art. This was defined, in the American case, in curiously Clausewitzian terms, and limited in scope to the purposeful linkage of battles and operations to achieve strategic purposes. With the elaboration of that concept, the doctrinal renaissance seemed to come to an end. This chapter will tell the story of how all that came to pass.

Doctrine, as a formal body of precepts, is not created in a vacuum. It is affected by institutional culture and experience. Although twentieth-century wars have been fought by the several armed forces acting in concert, the writing of service doctrine in the United States, prior to the Defense Reorganization Act of 1986, was a service function that might, under the best of circumstances, involve coordination between services but which left each organization generally free to go its own way within its own parochial view of the conduct of war. Army doctrine was written, its authors secure in the belief that its central goal was "winning the land battle."[3] This restricted view ensured that little attention was paid to more catholic and multiservice concerns, or to the divergence of perspectives on war fighting that divide the services, especially the ground and air forces. It led, inevitably, to a two-dimensional view of combat in which air-delivered weapons were tactical adjuncts to direct fire battles. Moreover, the limited scope of the problem tended to support a rather comfortable view that ground combat was the central activity in any conflict and any other players were simply supporting cast. These easy assumptions, seemingly compatible with the European focus of the Cold War, tended to limit, *a priori*, the set of possible solutions likely to emerge from any doctrinal speculation.

Aside from this limited view, it was the shattering experience of the loss of the war in Vietnam that conditioned the doctrinal revolution in the U.S. Army in the 1970s and 1980s. If the loss was not clear in 1973, when the last U.S. combat forces departed, it was self-evident to students at the army's Command and General Staff College in the spring of 1975. Vietnamese and Cambodian officer students did not go home that year. No amount of sophistry could conceal the fact that their dilemma was in large part due to U.S. failure in the war fought from 1965 to 1973.

Today it is almost impossible to gauge the effect of the loss of that war on the U.S. Army. The army disintegrated into an undisciplined organization without tactical or organizational standards. The leadership of the army had lost much of its moral authority, both in public debate on defense matters and internally. One general officer, after visiting the Staff College at Fort Leavenworth, reported to the army chief of staff that Staff College students were on the brink of mutiny.[4]

The army leadership identified the Soviet enemy in Europe as the only threat likely to merit a major military response (always excepting Korea, whose fortunes in American military calculations seemed to wax and wane depending on the administration in office). The Warsaw Pact armies were perceived to be an enemy that had used America's distraction to pull ahead in weapons quality as well as in numbers, the traditional Soviet advantage. Meanwhile, the Yom Kippur War of October 1973 demonstrated both the lethality and cost of modern warfare. The U.S. Army had to be refocused and retrained; discipline had to be restored; and an army that had missed a generation of weapons improvements had to be brought up to date. All of this had to happen at a time when the army's size was shrinking precipitously.[5] To lay the foundations for the redefinition and rebuilding of the army, the army chief of staff, General Creighton Abrams, ordered the creation of a new major army command, the U.S. Army Training and Doctrine Command (TRADOC). Abrams appointed General William E. DePuy as the Command's first commander.[6] Coincidentally, that same year the army abolished the Field Army as a relevant command echelon.[7] In NATO, operational headquarters were the army groups and they were alliance, not national bodies.

General Donn A. Starry, commandant of the Armor Center in TRADOC's early years, and subsequently General DePuy's successor, has observed that an army describes itself in four ways: in its doctrine, in the way it organizes, in the way it equips itself, and in the way it trains.[8] Training and Doctrine Command was charged with guiding all four activities. The first step in revitalizing the army was to be the development of a doctrine suitable for meeting the post-Vietnam threat, for refocusing an undisciplined army undergoing the transformation from a conscript to a volunteer force and changing it into an effective, highly trained, land combat organization. This was a task for which William E. DePuy was ideally suited. The circumstantial fact that Training and Doctrine Command was a new organization, asserting its authority over institutions accustomed historically nearly to independence in matters of doctrine development, and within an army unaccustomed to having its fighting doctrine prescribed with much authority, did not make his job any easier.

William E. DePuy was a decorated combat veteran of the Second World War, an experience that marked everything he was to do for the rest of his long and influential life.[9] DePuy was a tough-minded, highly perceptive, demanding commander. He had won the Distinguished Service Cross and a number of Silver Stars in the Second World War as a battalion commander and regimental and divisional staff officer in the Ninetieth Division. The Ninetieth was a division so inept in its first battles that General Bradley (or at least his staff) had considered its breakup.[10] The division consumed both commanders and soldiers to the point that DePuy later called it "A killing machine . . . of our own troops."[11] In Vietnam, General DePuy had commanded the army's First Infantry Division and served earlier as General William C. Westmoreland's operations officer.

DePuy was convinced that the army he was charged to reinvigorate and train was no longer a mobilization army but, rather, a force that had to be ready for combat at a moment's notice.[12] To train such a force required different standards than those demanded of an army with months available to achieve deployment standards. He was convinced time was short.

The need was self-evident. Aside from the internal problems visible on any army post, DePuy was greatly influenced by the lessons he learned from the 1973 Yom Kippur War; namely, that precision-guided munitions, wire-guided antitank missiles and new missile air defense systems had changed the nature of combat.[13] A favorite expression was, "If it moves it can be seen. If it can be seen it can be killed."[14] DePuy also retained admiration for the professionalism of his German enemy in the Second World War. He was very aware that America's German ally, alongside whom any U.S. defense in Europe must be conducted, would, in any event, have considerable influence on the acceptability of any new U.S. methods. Finally, DePuy was firmly convinced that of the four TRADOC functions—doctrine, organization, equipping, and training—the first among the equals was doctrine—the institutional expression of how the army would fight. Concept must lead action.

DePuy viewed doctrinal change as evolutionary, a point made clear early in his tenure in a letter sent to all school commandants in July 1974. The letter, which marked the beginning of the process of rewriting the Army's family of field manuals, began with a homely story:

> In France in the home of a peasant there is always a pot of soup boiling in the fireplace. From time to time someone throws in a potato, leek, some chicken stock or beef gravy, an occasional carrot or whatever. Over time the soup gets better and better. Everyone can add to it and anyone may partake. I view the attached paper somewhat in the same way.[15]

With the letter General DePuy forwarded a paper that addressed his views of how the army would fight in any future war.[16] The letter indicated an intention not to formalize the document and invited comment. DePuy wrote,

> Those parts of it which seem relevant and useful to your business should find their way into your doctrinal manuals and your instruction . . . and should provide a conceptual basis for the determination of weapons system requirements. Operational tests, force development test evaluations and experiments should be conducted in a manner consistent with the tactical concepts on which I hope we can agree through the medium of this paper.

DePuy indicated his desire that the paper would be a living document. "I don't care who sees it or how many copies are made. I just want to keep it like the pot of French soup."[17]

DePuy followed up the letter with a series of meetings with school commandants at Camp A. P. Hill, in Virginia, which addressed the development of a body of published doctrine for the army and undertook the writing of what came to be known as the capstone manual—FM 100-5, *Operations*. In the U.S. Army, Field Manual 100-5 is the lineal descendent of the first *Field Service Regulations*, published by a newly formed General Staff in 1905. The most famous edition was that published under the oversight of George C. Marshall, in draft in 1939, and in final form in January 1942. The 1976 FM 100-5 was a departure in almost every way from its predecessors. It was published in a loose-leaf notebook format, indicating an intention that it should be revised with regularity. It had a camouflaged cover to show it was a war fighter's manual. It was focused at battalion and brigade level, and it was even more narrowly focused, though DePuy sometimes denied it, to deal with the General Defense Plan problem in Germany.[18] There, political considerations and a long-standing reluctance to match Soviet masses with equal numbers obviated anything but a linear defense, and, indeed, a very thin one at that. More startling to traditionalists, the manual was dominated by numerous graphs, charts, and illustrations of the sort used by operations research systems analysts. These were intended to convince the reader of the lethality of the battlefield, the central reality from which DePuy and Starry deduced their doctrinal principles. This manner of argument proved unsettling after a war many professional soldiers perceived to have been lost due to application of Robert McNamara's quantitative management techniques to battle. The new manual prompted immediate controversy. Its novel format lent the appearance of being derived from quantitative formulae of the discredited McNamara "Whiz-kid" era. This combined with its prescriptive tone of detailed implementing instructions that followed in various subordinate documents (for example TRADOC Circular TC 7-24, *Antiarmor Tactics and Techniques for Mechanized Infantry* [30 September 1975]), more than its actual contents, bred almost immediate hostility to the new doctrine.[19]

Eventually Major General Donn A. Starry became General DePuy's principal confederate in developing FM 100-5, circumventing Major General John Cushman, commandant at Fort Leavenworth. Cushman wanted a more traditional narrative and abstract approach to the emerging doctrine than did DePuy, but DePuy was in too great a hurry to convince his subordinate at the Staff College.[20] In fairness to the TRADOC commander, the Staff College draft suffered from the lack of experience in writing doctrine that then obtained in the army.[21] Disappointed with the rate of progress in preparing a new operations manual, DePuy had some of his trusted general officers write draft chapters at a subsequent meeting at A. P. Hill. He then took the draft back to TRADOC Headquarters at Fort Monroe, Virginia, formed a cell of writers—composed mostly of young field grade officers—and, with Starry's collaboration, produced what was to be the most controversial manual in the history of army doctrine.[22] It proved to be a manual greatly underestimated both in terms of its perception and its long-term influence. The draft became the 1976 FM 100-5, *Operations*.

The heart of FM 100-5 was something called "active defense," though the manual itself says very little about it. Active defense, which apparently was created by General Starry at Fort Knox,[23] was essentially a defensive tactic for brigades and battalions. It called for a commander, subject to an enemy's principal effort, to shift forces from less committed sectors to the point of main threat in a process sometimes called *thickening*. Despite the manual's chapter on the offense, which spoke of deep penetrations and envelopment, and support for the idea that leaders needed to be left free to act within a superior commander's general intentions, FM 100-5 appeared to support an attrition-based response.[24] Success was based on creating a favorable balance of forces along the battle line as a means of dealing with an enemy penetration—of bending rather than breaking. Counterattack was viewed as essential but discouraged in most cases as too risky. Overall, the tone of the manual was decidedly pessimistic.[25]

Active defense was rightly criticized as attempting to match strength with strength while surrendering the initiative to a vastly superior force. Still, limited as it was to consideration of brigades and battalions, it was not a bad picture of what the tactical commander in Europe was likely to confront. Faced with a Soviet breakthrough attack, a brigade commander was going to be saddled with an active defense simply by virtue of the number of enemy forces confronting him and the tempo of assaults to which he would be subjected. His brigade might die more or less elegantly—but it would die. Its success would be measured by the time it bought for some other commander, higher up the chain of command, to find an operational mass of maneuver, or to decide where and when to employ nuclear weapons—the latter as much for political or moral effect as any purely military utility. FM 100-5 said nothing of note about that part of the battle, however, and a good deal of the subsequent criticism was that the manual was a recipe for failure. FM 100-5 was rushed into publication without wide staff review in the army.[26] Within three years, it would be revised.

Equally important, and more long-lasting than the prescriptions of the 1976 FM 100-5, was the idea that accompanied it, *that it should be followed.* American soldiers have long prided themselves on improvisation—not tactical discipline. DePuy and his second principal subordinate, then Brigadier General Paul Gorman, one of DePuy's principal subordinates in the First Infantry Division in Vietnam, set out to change that too—to restore tactical discipline by requiring compliance with standards of field craft, tactical principles, and even such minor tasks as the proper way to debus an armored vehicle. Evaluated training to standard became the army way.

First, Gorman began a process to produce a Cartesian analysis for every combat task (i.e., to reduce each task to its simplest components) and then to establish standards of performance against which their execution could be evaluated.[27] Gorman and DePuy drew upon new laser technologies to lay a foundation for evaluated free-play exercises and, ultimately, to create the National Training Center in California. Although the actual conduct of training remained the province of unit commanders, Training and Doctrine Command established the standards and methodologies that govern army training to this day.

In June, 1977, William E. DePuy retired from the army. Much that he had written would be subjected to criticism and revision. Building doctrine from the bottom up, in terms of the long view, may have gotten the problem exactly backward; however, its role in restoring discipline to army training is sometimes overlooked by critics. As Donn Starry has said recently,

> We had to do something. We probably could have done almost anything as long as it was different and everybody would have seized on it. But AirLand Battle [which succeeded Active Defense] was the end result of about ten years of working hard on doctrine, on equipment requirements, on organization, on training, and on the education of officers, NCOs, and soldiers. . . . That, I think is Gen. Bill DePuy's greatest legacy to all of us. He had a conception in which all that fit together.[28]

General DePuy was succeeded at TRADOC by General Starry in the summer of 1977. By then a counterrevolution in doctrinal development was emerging, fueled by two principal factors. First, the army in the field set about trying to learn the new tactics embodied or extrapolated from the new manual.[29] In the Eighty-second Airborne Division, for example, commanders trained light infantry battalions using an interim manual for heavy infantry, to learn variations of the active defense called "the Airborne Anti Armor Defense."[30] Units made their own lists of tactical tasks, conditions, and standards, to begin testing themselves in accordance with General Gorman's principles while awaiting the publication of what remained a paper-based training system.

More important, as it would turn out, General Starry went to Germany to take over as Fifth Corps commander. There he became the "outside man" in the DePuy-Starry team, trying out the tactical concepts on the ground for which they had been designed, and discussing them with his German ally.[31] Starry quickly determined that there were serious shortcomings in the doctrine; it was inadequate to the task of stopping a Soviet breakthrough attack unless a means could be discovered to better meter the arrival of new enemy units at the friendly line of contact.[32] Starry's experience in Germany would have a profound effect on the next stage of doctrinal development when he was called upon to succeed General DePuy at TRADOC in 1977. It is important to remember that, from this point on, intellectual speculation was accompanied always by empirical tests conducted regularly by units in the field employing the DePuy-Gorman vision of training accountability, something DePuy had not had available to him when he started his doctrinal revival. It is also important to note that while a discussion of the operational and strategic implications of NATO's forward defense had begun in the civilian defense community, the army's focus remained essentially tactical.

At the same time officers in the field were trying out the tactical techniques of FM 100-5 and its subordinate publications, it became apparent to many that there were certain internal contradictions in the 1976 manual. First, the "capstone" manual was titled *Operations*, but it was, quite obviously, a tactical manual, and a minor tactical manual at that. Second, it was theater specific and relatively passive, or, at least, reactive, a posture suitable for the political requirements along the inter-German border but not acceptable for the doctrine of an army with global responsibilities, or even for the cultural values of the American officer corps. Moreover, as one relatively young officer pointed out in an article written while FM 100-5 was in draft, the new doctrine seemed to be so concerned with the survivability of friendly units in small unit battles that it lost sight altogether of the needs of higher-level commanders.[33] Another field grade officer writing in the Staff College journal, *Military Review*, in 1978—the same year Starry laid out his own concerns about the new doctrine—noted that the army had lost sight of the need to address "large unit operational doctrine."[34] These two essays, however, remained exceptional in a growing controversy that, like the manual itself, tended to remain tactical in its focus.

There was also a stream of lively criticism stimulated by various civilians. Lacking popular legitimacy due to the loss of the Vietnam War, the Army hierarchy was subject to widespread questioning. Congressmen, academics, congressional staffers, and even Staff College students, felt themselves competent to criticize the generals who were viewed as responsible, legitimately or not, for the fortunes of the postwar army. "Pot of soup" or not, this cacophony did not wear well on strong-willed men like DePuy or Starry.[35] Nonetheless, shortly after publication of FM 100-5 in July 1976, a congressional staffer named Bill Lind broke the dike inhibiting criticism of the new doctrine, and opened thereby an important debate on the army's fundamental beliefs about battle. In 1977 Lind published a critique of the manual in *Military Review*[36] which infuriated the army leadership.[37] It also made it legitimate for anyone to question the capstone doctrine. Lind's lead was followed by a real debate in the pages of *Military Review* and *Army* magazine. It caught the thinking army's attention and sparked a continuous and fruitful interchange of ideas that would have been unheard of ten years earlier—or today, for that matter.

Lind's article was typical of much of the criticism of the DePuy-Starry doctrine. He offered a polar taxonomy of war featuring, what he called maneuver warfare, opposed to what he labeled firepower attrition, and he placed the new army doctrine firmly in the latter camp. He argued that firepower attrition was unlikely to succeed in a European context, that overemphasis on winning the first battle could leave a defending force too weak to fight the second, and he championed uncritically the use of maneuver, "as a weapon in itself."[38] Lind used historical examples taken from Second World War German offensives and more recent (1973) Israeli counteroffensives, and questioned, as did many army officers, whether the use of uncommitted flanking forces was a realistic alternative to maintenance of more traditional reserves. However, Lind and many like-minded critics failed to acknowledge the necessity of adapting tactical doctrine to strategic requirements. He asked at one point whether the new doctrine might "possibly constitute the sacrifice of a militarily correct principle that depth is desirable in the defense, to a political requirement,"[39] specifically, placating America's German ally.

But Lind failed to address how offensive tactical maneuver could be better accommodated within the NATO framework. That is, he failed to frame his question within the conditions to which the new doctrine was required to conform, and, thus, he ended up talking past those actually trying to solve the problem. Implicit in Lind's criticism, though generally lost in the attrition-maneuver dichotomy, was the question of whether tactics, a subordinate and limited activity, should be dealt with in terms of organic context, rather than as an ideal type, completely independent of limitations of context and purpose. Lind clearly chose the latter.

Serving officers also joined the public debate. Two of the most significant authors to address the need for more traditional and idealized "maneuver" doctrine were then Colonel Wayne A. Downing, later a four star general, and Colonel Robert E. Wagner, head of the Second Armored Cavalry Regiment when he wrote "Active Defense and All That," who went on to become a major general. Downing gave Lind's ideas increased legitimacy and a clearer articulation.[40] Wagner, reputed to be one of the army's premier tacticians, demonstrated that within a NATO defensive zone, in his case a regimental sector in Germany, there were opportunities to practice more aggressive defensive options using terrain, obstacles, fire, and concentrated reserves.[41]

An even more noteworthy, and certainly more balanced, article was written by then Major Richard Hart Sinnreich. Sinnreich would one day be a director of the army's School of Advanced Military Studies and author of a future FM 100-5. When he wrote his essay "Tactical Doctrine or Dogma?" in 1979, he was serving at the Field Artillery School.[42] Sinnreich laid the arguments of the maneuver warfare critics against the European context and criticized both the critics and the doctrine with clarity and precision. He pointed out the various confusions that had grown up in the brief life of the new doctrine, attributing much of it to various imprecisions in language and concept. Not least important, he pointed out that doctrine addressing actions of brigades and battalions was somehow being generalized into prescriptions for larger unit actions. He concluded, "The point is that tactical doctrine is neither autonomous nor absolute. It can be divorced neither from the tactical circumstances in which it must be applied nor the strategic purposes it is intended to further."[43] The essay closed with the observation that the institutional army was reacting to criticism by hardening its insistence on particular and perhaps inappropriate formulas.

Sinnreich's essay won an approving response from then retired General William E. DePuy.[44] The following year, 1980, DePuy himself admitted it was probably time to revise the doctrine he and Donn Starry had developed.[45] DePuy observed, in *Army* (November 1980), that attention to "active defense," particularly in subordinate manuals, had served to crowd out consideration of other tactical options, and he acknowledged the need for a body of diversified tactics that could be adopted in accordance with circumstances. He insisted on the importance of the principle of elasticity, whatever options were selected, pointing out that,

> Part of the problem with criticism of active defense surely stems from the deep frustration associated with the defensive strategy adopted by the NATO alliance. The decision to defend is the most the NATO commanders believe they can extract from the forces available. *These are strategic and operational rather than doctrinal or tactical decisions.* [Emphasis added.][46]

DePuy's use of the term "operational" along with "strategic" and "tactical" is indicative of a growth in conceptual framing that would have a significant influence on the direction of army doctrine. It is one indication that thoughtful officers were starting to look for a way to find expression for a level of activity that would connect tactical actions and strategic purposes.

By the time these officers wrote their essays, FM 100-5 was already under revision. The effort began as a result of a discussion between General Edward C. Meyer, in June 1979 the army's deputy chief of staff for operations and new chief of staff designee, and his subordinate, Major General William R. Richardson, the assistant deputy chief of staff for operations, force development. Meyer's principal concern seems, in light of Richardson's response, to have been the narrow focus of the 1976 manual, particularly the failure to consider minor tactics within the framework of larger unit operations.[47] Richardson provided Meyer with a draft letter to General Starry that recommended a revision of the manual in light of three weaknesses: limitations of scope, threat, and balance.

Meyer's letter observed that the European focus had been reasonable in 1976 but noted that, while it confronted the most important security threat, that threat was also the least likely to be realized. "We need to expand on this theme," he wrote, "and address the other wars as thoroughly as we have already treated the Central European case."[48] The revised manual must consider corps and theater battlefields, "where their managers must execute the strategy and perform those essential executive functions that provide overall direction to the effort and which insure that resources are efficiently brokered."[49] Consideration was to be given enemy options other than the breakthrough and something was to be done to mitigate the defensive focus that seemed to permeate the manual. "Although we may often find ourselves in a defensive mode at varying levels of echelons," Meyer continued, "we still expect and intend to carry the fight to the enemy. The opportunities will vary; but as a matter of doctrinal procedure we will customarily think and operate in an offensive mode."[50] Anticipating that such a letter might ruffle the TRADOC commander's sensitivities, Meyer paid due homage to the accomplishments of the drafters of the 1976 manual, not just for addressing the complex and largely impossible task in Europe in the context of the future rather than the past, but for instigating the doctrinal debate that followed. "It has caused people to think aloud for a change."[51]

Starry's response indicates some irritation, though he set to work at once doing what Meyer had suggested.[52] The TRADOC commander acknowledged that the 1976 manual had addressed what he and DePuy considered the most challenging problem, one it was necessary to address to rationalize U.S. and German doctrine for NATO. He admitted that they had been unable to find an adequate response to the echeloned nature of a Soviet attack. "We know what needs to be done about that now," he wrote, "and so can write coherent doctrine that will cover both offense and defense, and the fight against the first echelons and the second echelons as well."[53] Then, he added with obvious reference to the maneuver warfare theorists' advice for European defense:

> However, our critics who trumpet that we should steer away from a tripwire based strategy aren't all correct . . . we're going to have to kill a whole lot of them [Russians]—just to get their attention. And we should make no mistake about that! So to say that it's all a war of maneuver and that maneuver will solve all, is to ignore the very real problems with space and depth, especially in Europe and with logistical support of highly mobile operations. For certain we must find some prose that's more golden than that we used last time.[54]

Starry, then, began what to him was essentially a revision of the existing manual rather than writing a new one.[55] He sought to solve the problem of echelonment by assigning responsibility for the direct or close battle to the divisions, brigades, and battalions. The corps would fight what came to be called the deep battle, with indirect fire weapons, organic aviation, and air force strike assets. The concept governing this division of the battlefield was called the "extended battle."[56] It reflected what Starry had learned as the Fifth Corps Commander in Europe. Starry and Richardson eventually decided to combine the two simultaneous fights under the name "AirLand Battle," a term first used in the 1976 manual to describe the close air support problem.[57]

Bill Richardson, the officer who had stimulated, or at least given focus to the need for revision, was reassigned to Fort Leavenworth as commander, Combined Arms Center, and commandant of the Command and General Staff College. Richardson provided an interesting contrast to his new commander, Donn Starry. Starry is a big and powerful man to whom great authority seems natural. A veteran of the Korean War, commander of the Eleventh Armored Cavalry Regiment in Vietnam, Starry went on after retirement to enjoy success in the business world.[58] Starry, like DePuy, did not tolerate fools, or even apparent fools, gladly. Widely read, Starry forced history into the Staff College curriculum and faculty, not without resistance.[59] He was largely responsible for the tactical doctrine imposed on the army in 1976, and in 1979, he clearly intended to play an equally active role in whatever doctrinal revisions were necessary. Richardson, in contrast, is a quiet intellectual, almost courtly. As commandant at Leavenworth, then deputy chief of staff for operations, and finally as Starry's successor-but-one at TRADOC, Richardson provided continuity of leadership for the army's doctrine and related educational programs over the next seven years, a period as long and fruitful as that of DePuy and Starry before him. Though Richardson's direct influence on army doctrine is more difficult to pin down, and less personal, his decisions were decisive in setting the course of army doctrine as it changed from something developed in splendid isolation by senior warlords at TRADOC to the product of a more collective and bureaucratic process, centered at the Staff College, and underpinned by the intellectual training and speculation going on there.

This occurred in the first place because General Starry decided to return doctrine writing to the TRADOC schools.[60] He did so because he believed that if they did not write the doctrine, the school faculties could not explain it properly, and students would leave the schools misinformed about army doctrine. Starry recognized that the schools could be a valuable mechanism for changing the army's way of thinking. The Staff College released about 800 mid-career officers to the Army's mainstream annually. Richardson shared this view and does so to this day.

Richardson assigned responsibility for the revision of FM 100-5 to three lieutenant colonels, nominally assigned to the College's department of tactics but actually working directly for the commandant and TRADOC commander, Richardson and Starry. The first of these was Lieutenant Colonel Richmond B. Henriques. Henriques retired from the army in 1982 after the new version of FM 100-5 was published. He was joined in the summer of 1980 by Lieutenant Colonel Huba Wass de Czege, an infantry officer, West Point graduate, Harvard-trained political scientist, and former instructor in West Point's Social Science Department. Wass de Czege would head the revision effort. The third officer was Lieutenant Colonel Leonard "Don" Holder, a graduate of Texas A&M University, a cavalryman, and former instructor in the Military Academy's Department of History.

Richardson had known Wass de Czege at West Point when the latter was a cadet. Holder was a protégé of Colonel Bob Wagner. They formed a remarkably complementary team. Wass de Czege was a charismatic Hungarian romantic, Holder a laconic and highly practical Texan. Together, Wass de Czege and Holder would dominate much of the creative effort in the evolution of army operational doctrine during the next six years, collaborating under Richardson's oversight on two editions of FM 100-5, and overseeing the development of the concept of operational art as a doctrinal construct. Influenced by the classical and modern theorists of maneuver warfare, Holder and Wass de Czege would contribute a conceptual broadening to army doctrine essential to making that doctrine an interpretative as much as a prescriptive body of knowledge.[61] Their revised manual recast the old DePuy-Starry active defense concept, reverting to more traditional modes of description. The analytic charts and graphs were deleted. They added historical allusion, restored the nine traditional principles of war, and, for those who found nine too challenging, added four tenets, or characteristics of combat: agility, depth, synchronization, and initiative.[62] In fact the 1982 manual was something of a book of lists. There were four tenets, three levels of war, four elements of combat power, seven combat imperatives, and so on.

In later years, after the final draft was approved by Starry's successor at TRADOC, General Glenn K. Otis, Wass de Czege and Holder would downplay Starry's influence on the revised manual.[63] They could point to the concept of initiative as the principal element of the new doctrine as opposed to Starry's idea of the extended battle. "The object of all operations is to destroy the opposing force," the new manual read.

> The Army's basic operational concept is called AirLand Battle doctrine. This doctrine is based upon securing or retaining the initiative and exercising it aggressively to defeat the enemy. Destruction of the opposing force is achieved by throwing the enemy off balance with powerful initial blows from unexpected directions and then following up rapidly to prevent his recovery. The best results are obtained with initial blows struck against critical units and areas whose loss will degrade the coherence of enemy operations....[64]

Aside from an obvious offensive tone, the manual restored a more traditional taxonomy or classification scheme for types of combat operations. Its basic framework was more comprehensive. Even so, Starry's practical concepts were present in the document and, as his letter to General Meyer showed, Starry already believed in force-oriented combat.[65]

General Richardson made clear his understanding of the importance of indoctrinating Staff College students in both the extended battle and the integrated battlefield (a battlefield on which weapons of mass destruction could appear without warning), as he showed in an article written for *Army* to accompany the new manual and in a tour-end report to General Starry.[66] The manual was undoubtedly written at the Staff College, but it did not become doctrine until approved by the TRADOC commander and army chief of staff.

The real difference between the 1982 FM 100-5 and its 1976 predecessor was the extent to which the revisionists were able to change the approach and tone. More traditional and culturally acceptable narrative methods of description were used, and prescriptive admonitions were set within a larger conceptual pattern. The descriptive and prescriptive had been rebalanced, though in fairness to the authors of the previous (1976) edition, much of the prescription could now be borne by subordinate manuals written since 1976, thus permitting the capstone manual to assume a more abstract tone, to be more of a textbook than a pocket guide.

In the summer of 1981, when the new manual was about ready to go to press and after wide circulation to the army at large, Starry was reassigned to U.S. Readiness Command and replaced by General Glenn K. Otis. Otis delayed publication of the manual, insisting that certain ideas be clarified. He was particularly concerned that what he saw as the close, equal, and symbiotic relationship between fire and maneuver be clearly articulated.[67] Otis also directed that the concept of "operational level of war" be added to the doctrine written by Wass de Czege and Holder.

The initiative for this addition came from Brigadier General Don Morelli, the deputy chief of staff for doctrine at TRADOC. Morelli credited the Army War College for developing the concept. Indeed, the College was the site of another ongoing intellectual revolution led largely by Colonels Harry Summers, Wallace P. Franz, and Arthur Lykke. Franz was prominent among those who pushed German operational concepts and terms into the army's consciousness.[68] Lykke, the professor of military strategy, published the War College text, *Military Strategy: Theory and Application*, and indoctrinated a generation of senior officers into thinking of war fighting in terms of *ends, ways, and means*.[69] Summers, aside from writing an important analysis of the Vietnam War, or perhaps because of it, was perhaps the major influence on the U.S. Army's turn toward Clausewitz as a theoretical touchstone.

Although the principal Leavenworth authors resisted the addition of the "operational level of war," arguing that the concept was too difficult for the army to grasp, the concept eventually found its way into the army's capstone doctrine.[70] The operational *level* was defined imprecisely as a "broad division of activity in preparing for and conducting war."[71] This decision to view war as a set of "levels," with the implication of place rather than categories of action, produced some definitional awkwardness. Rather than defining operational art, the manual simply made assertions about what things would fit in an operational level. With a surprising indifference to antecedent, the authors wrote that the "operational level of war uses available military resources to attain strategic goals within a theater of war." They then asserted that "most simply, it is the theory of larger unit operations." Most important, though, it introduced into the army's capstone war-fighting manual the concept of the *campaign*. "Campaigns," the manual went on, "are sustained operations designed to defeat an enemy force in a specified space and time with simultaneous and sequential battles."[72] An army that, heretofore, had occupied itself almost exclusively in the preparation for fighting battles now turned its attention to the creative articulation of the battles in the ensemble.

On a trip to China with Richardson—taken while Don Holder wrote the final draft of the revised manual—Wass de Czege got approval to create a new school at the Command and General Staff College to study large unit operations, and, by implication, seek a better understanding of "the operational level of war." Wass de Czege spent the following year (1982-1983) as a War College fellow at Fort Leavenworth, pursuing an independent course of study, sponsored by the new army deputy chief of staff for operations, William R. Richardson. Wass de Czege's purpose was to develop a curriculum for a course focused on large unit operations and specifically the operational art.[73]

The new School of Advanced Military Studies (SAMS) of the Command and General Staff College accepted its first students in June 1983. Colonel Huba

Wass de Czege was it first director. Don Holder would be its third. Wass de Czege hired a young analyst named Jim Schneider to teach the school's military theory course. Schneider combined the talents of a systems analyst with an undergraduate background in philosophy and a graduate degree in history. (He later completed his Ph.D. in Russian history, writing his dissertation on the development of Russian military theories of operational art.)[74] SAMS students studied classical theory, principally Clausewitz's *On War*, and examined large unit operations in history and in simulations, in order to understand what the school came to call the operational art. There was something of an air of conspiracy about SAMS in the early days. It was breaking new doctrinal ground, and each year its graduates went to key planning positions in the army's divisions and corps, spreading the SAMS gospel. The motto selected to describe the SAMS graduate, attributed to the great Graf von Schlieffen, was revealing: "Work Relentlessly, Accomplish Much, Remain in the Background, and Be More Than You Seem."[75]

In September 1984 Wass de Czege, still director of SAMS, was called in by General Richardson, now TRADOC commander, and told to undertake a new revision of FM 100-5. Before examining why a third revision was necessary, it is useful, first, to reflect on the development of the concept of an operational level of war and its correlate, operational art. Where had the idea come from?

Perhaps the most likely source is Edward Luttwak, the well-known defense critic. Luttwak had published an article titled "The Operational Level of War" in *International Security* in the winter of 1980-1981. The article was a reworking of a paper written for Andy Marshall's Office of Net Assessment in March 1980, titled "National Styles in Warfare and the Operational Level of Planning, Conduct and Analysis."[76] Luttwak pointed to the absence in English of a word to cover the dimension between tactics, or the conduct of battle, and theater strategy, for the allocation of resources against political goals. In short, what was missing was a governing concept for those activities which employ tactical events in a purposeful way to achieve the goals of strategy. Offering a somewhat tendentious historical analysis, Luttwak asserted of modern American military practice, with particular reference to the Second World War, that "above the purely tactical level, the important decisions were primarily of logistic character. The overall supply dictated the rate of advance, while its distribution would set the vectors of the advancing front."[77]

Luttwak argued in effect that lacking a term for operational-level practice, U.S., and presumably British, and Commonwealth armies, also suffered from an absence of operational thinking. Notably, Luttwak concluded his essay by observing that "it is also true that the politically imposed theater strategy of Forward Defense precludes the adoption of the only operational methods that would offer some opportunity to prevail over a materially more powerful enemy."[78] In short, he disqualified his theoretical premise for strategic unsuitability when confronted with practical realities. In fact, he had hit upon an idea already percolating in the army.

Many in the army were aware of the concept of the operational level. In the 1970s there had been a strong interest in Soviet doctrine. The air force was translating and publishing a number of Soviet theoretical works, among them V. Ye. Savkin's *Basic Principles of Operational Art and Tactics*.[79] General DePuy had insisted on the need to study the Soviet enemy, and the army had taken him at his word. It was apparent that the Soviet Army did have a coherent and comprehensive body of military theory that was worthy of attention. Indeed, emerging U.S. doctrine sometimes suffered by comparison to the threat manuals turned out by the intelligence community.

At the same time, the first postwar analyses of the Vietnam War were creating a view, later articulated by Harry Summers,[80] that something had been missing from U.S. conduct in that conflict. At the Army War College, Colonel Summers was turning officers' attention to Clausewitz. The Peter Paret and Michael Howard translation of *On War* appeared in 1976, and those who assayed the Prussian soldier-philosopher found the Clausewitzian definition of strategy, "the use of engagements for the object of the war,"[81] a source of disquiet about the conduct of the late conflict in Southeast Asia. Howard himself pointed out the inadequacy of the signifier "strategy" in his seminal *Foreign Affairs* essay, "The Forgotten Dimensions of Strategy."[82] The word, "strategy," might no longer do, but the concept Clausewitz wrote about, the purposeful use of combats, was everywhere in the air, call it campaign strategy or whatever.

It had become increasingly clear, then, that the real artistry of war took place on a plain of action above the winning of a single battle. Battle was a means, not an end, in war. Moreover, there was a dimension of action outside the practical experience of most officers, which might have its own logic and grammar, and it might be mastered by study and reflection. It was clear from a briefing prepared for General Otis at TRADOC Headquarters, in the spring preceding the publication of the new FM 100-5, that the army's senior leadership had a pretty good idea of what they thought the operational level of war was, even if they could not yet define it with precision.[83] The leading characteristic of actions at the operational level were identified as defeating the enemy by the relative positioning of friendly forces (maneuver), through the conduct of campaigns, comprehending the *theater of operations*.[84] The rationale for adopting the idea of levels of war was to instruct senior commanders to differentiate between the variable natures of fundamental categories—specifically of maneuver—at each level, and to explore the interrelationships that existed between the levels themselves.[85] Notably, under a briefing slide titled "Strategic Level" was the assertion, "No coherent strategy in Viet Nam."[86] What was proposed with this particular innovation was no less than the adoption into doctrine of a holistic and integrated view of warfare, one in which the operational level was to comprehend large unit actions within a theater of war.

The discovery, perhaps the rediscovery, of operational art—the creative activity practiced at the operational level—was therefore a by-product of trying to understand the American loss in Vietnam. The operational level, the operational art, not only pointed forward to preparation for the next war, but provided a certain retrospective closure to what was increasingly seen, rightly or wrongly, as the general bankruptcy of the higher U.S. conduct of military operations in the lost cause of Vietnam,[87] as well as the operational futility of forward defense as a war-fighting method. That made the distinction, and thus the connection, between simply winning battles and winning wars.

It is important to acknowledge the role played by Ed Luttwak in shaping the revised doctrine published in 1982. His 1980-1981 article had introduced into public debate the concept of the operational level of war—a concept to which the army was particularly receptive. Luttwak was also one of the many defense critics consulted by army leaders when, in an effort to win back the institution's professional authority, they marketed the new doctrine to the interested public.[88]

The next stage in the growth of AirLand Battle—namely, the preparation of a final, pre-Desert Storm manual in 1986—was dominated by an attempt to redefine the operational level of war. Beginning in 1984 at the School of Advanced Military Studies, the redrafting was overseen by Wass de Czege and assisted by Holder (who came to SAMS as a War College fellow) and Lieutenant Colonel Richard Hart Sinnreich, Wass de Czege's deputy at SAMS. Sinnreich, who succeeded his superior when both Wass de Czege and Holder left in the summer of 1985, would eventually write the publication draft of the 1986 FM 100-5.

Sinnreich possessed an extraordinarily sharp intellect and a wonderful talent for precise writing developed under the tutelage of Dr. Henry Kissinger. He had been involved in monitoring the army's adoption of AirLand Battle doctrine, both on the Army Staff for General Richardson, while Richardson was deputy chief of staff for operations, and later at SAMS.[89] Moreover, Sinnreich had worked for the Supreme Allied Commander Europe (SACEUR) at his Headquarters and on the National Security Council Staff. This experience gave him an unusual sense of the politics of service doctrine, particularly from the European perspective. That was important precisely because, while the internal cause of the 1986 revision was the elaboration of theater-level warfare and operational art, an external reason was the continued hostile reception to army doctrine in Europe, first by Alexander Haig, for whom Sinnreich had worked, then by General Bernard Rogers.

It can be argued that army doctrine was not in tune with changing European defense opinion. The 1976 doctrine, considered defeatist by many in the U.S. Army, was published at a time when there was great concern in NATO both about U.S. commitment and NATO capabilities.[90] The Nunn-Bartlett report, which led to the start of what has been mislabeled the Reagan defense buildup, was released in 1977.[91] Then, the 1982 manual, with its prescription for deep attack, was released at a time when Europeans were becoming alarmed about the new U.S. president's talk of "evil empires." The new AirLand Battle doctrine seemed unduly offensive to some.[92] Moreover, General Rogers had his own

NATO doctrine called "Follow-On Forces Attack."[93] That this seemed to differ little from the extended battlefield, except in language, did not relieve the friction between the army and its senior general officer.

Wass de Czege pointed to these two requirements when he sent the new draft manual to the field in July 1985.[94] He argued that the fundamental doctrine contained in the 1982 manual had been accepted by the army and pointed out that a good deal had been learned since then about the operational level of war. The principal difference between the 1982 and 1986 manuals was the extent to which the new volume addressed campaigns and sustained multi-engagement operations. Wass de Czege had always considered the attention given the human dimension of combat a fundamental shift in emphasis in the 1982 manual. This, he indicated, had been retained. Finally, he observed that AirLand Battle doctrine had been subject to a number of misinterpretations. "The gravest of these," he now wrote, "has been that AirLand Battle is a *strategy*."[95] The new manual had been carefully worded to ensure that any reader would understand that victory, or success in any war, was defined not by the logic of tactics but by the needs of policy.

A separate paragraph of his letter addressed specific concerns in Europe. Notably, Wass de Czege observed that considerable effort had been made to harmonize the air force's *a priori* theater-level view with the desire of army corps commanders for reliable air force assets in support of their own activities. Wass de Czege pointed out the obvious: corps commanders could not assume they would receive air force support independent of theater priorities, a point which may have been accepted philosophically but which remained a highly emotional issue for the army's corps commanders in the Gulf War, regardless of army doctrine.

The new manual drew on four years of student studies at the School of Advanced Military Studies, in which Wass de Czege and Holder took part.[96] These were influenced by work at the Army War College, most notably international seminars on the Soviet-German war of 1940-1945, conducted annually by Colonel Dave Glantz.[97] The Soviet influence was also felt through the efforts of the Soviet Army Studies Offices, later the Foreign Military Studies Office, founded at Fort Leavenworth by Dr. Bruce Menning in 1985, and later headed by Colonel Glantz.[98] The SAMS course was defined largely by historical studies of modern, and usually armored, campaigns and analyzed in terms of Clausewitzian concepts, particularly *centers of gravity* and *culmination*. The exemplars of operational art were taken to be U.S. General George S. Patton, and the Germans, Field Marshal Erwin Rommel, General Heinz Guderian, and General Erich Von Manstein. Jim Schneider added theoretical readings by Soviet Colonel V. K. Triandafillov and Marshal M. N. Tukhachevsky. The trend was clearly to define an operational art in Clausewitzian terms derived from large continental wars, whatever the likelihood of such conflicts (though one must remember that, a decade ago, the unlikelihood of European war was by no means considered certain).

The final manual of the series considered here was published in July 1986. It is much better written than either of its predecessors. General Rogers's sensitivities had been handled very carefully by General Richardson at every step of the process.[99] Indeed, the preface to the manual noted that "FM 100-5 is compatible with and will serve as the US implementing doctrine for NATO land forces tactical doctrine (Allied Tactical Publication 35A)," and accepted explicitly that U.S. Army forces in NATO would execute NATO's forward defense plans in compliance with NATO doctrine.[100] The second chapter, which was the condensed expression of the doctrine, now spoke of the "Structure of Modern Warfare," consisting of military strategy, operational art, and tactics, "the broad divisions of activity in preparing for and war."[101] The object of operational art was given as translating the goals of policy into effective military operations and campaigns. Operational art was defined, finally, as "the employment of military forces to attain strategic goals in a theater of war or theater of operations, through the design, organization, and conduct of campaigns and major operations."[102] Operational art, the chapter went on, "thus involves fundamental decisions about when and where to fight and whether to accept or decline battle. Its essence is the identification of the enemy's operational center of gravity. . . and the concentration of superior combat power against that point to achieve a decisive success." Three questions were posed for the operational planner:[103]

(1) What military conditions must be produced in the theater of war or operations to achieve the strategic goal?
(2) What sequence of actions is most likely to produce that combination?
(3) How should the resources of the force be applied to accomplish that sequence of actions?

With that, the evolution of the doctrinal void was filled and the circle of theoretical and doctrinal development was closed. So, more or less, was the intellectual storm which had followed the army's attempt at doctrine writing in the post-Vietnam era.

CONCLUSIONS

What is most striking about the United States's adoption of the concept of operational art is that the process was almost entirely synthetic, abstract, and imitative. Unlike the continental powers, which seem to have evolved concepts of operational art during or in anticipation of the conduct of extensive land campaigns, the United States came to the idea through the study of military operations of a type divorced from those they seemed at all likely to undertake. Coming out of Vietnam, the U.S. Army was overcome by a sort of professional nostalgia, a yearning for the simpler days of Napoleon, or Grant and Lee, or even Patton (if one limited one's view to Patton's pursuit to the Seine, or Third Army's counterattack into the Bulge, and did not concern oneself unduly with such things as the Seigfried Line operations). At the School of Advanced

Military Studies students were also fascinated with the practitioners of German blitzkrieg.

That the idealized models were in some way not appropriate, however, does not mean that they were not useful. Clausewitz himself found out that it is far easier to write a theory of idealized war than to address its reality, constrained by circumstances, conditions, and considerations of cost and benefit.[104] War *in general* is simply far easier to write about than war in particular. The distinction is not one with which American army officers are comfortable. Moreover, the concepts developed thinking idealistically are useful for ordering one's thoughts and dialogue about real problems. The idea of *center of gravity*, for example, has proven an excellent tool for focusing one's efforts on more valuable objectives; *culmination* reminds operational planners of the tyranny of logistics, no less the disorganizing effect of victory,[105] and *operational art* introduced to an army fundamentally concerned with tactical questions the notion that tactical events were simply a part of a larger strategic whole. The latter development addressed theoretically the criticism of Summers's North Vietnamese colonel, who complained that all the battles the United States won in Vietnam had, ultimately, not contributed one whit to the accomplishment of America's strategic objectives.[106]

During the period between the collapse in Vietnam and the Gulf War, army doctrine, most particularly the incorporation of the idea of operational art, was negotiated more than realized as the product of some divine inspiration. It is arguable that, had General Glenn Otis not had the determination to settle the maneuver-firepower debate, the issue of the levels of war, with different manifestations of concepts, would never have arisen in 1981. Had it not, the concept might have developed along different lines altogether. It would certainly have come later. The evolution of doctrine was another case of C. P. Snow's observations on the development of government policy: "shaped under one's eyes by a series of small decisions.... [Built] from a thousand small arrangements, ideas, compromises, bits of give-and-take."[107]

The idea of operational art did fill a significant void in U.S. military thought. It provided a concept to relate tactical events to strategic outcomes and provided a framework within which to think, even tentatively, about large unit operations. It also offered an arena in which tactically focused army generals could talk with their more operationally attuned air force counterparts on the relative value of particular tactical events in terms of a common calculus, though it is not altogether clear that they have done so.[108] That too little of this debate and discussion still goes on is, perhaps, indicative of the need to continue pressing for further development of the operational art concept in an armed forces once more caught up in a perceived technology-based revolution in military affairs.

NOTES

1. The views expressed in this chapter are those of the author and do not reflect the official policy or position of the Department of Defense or the U.S. Government.

2. Based on the story related in Colonel Harry Summers, *On Strategy: A Critical Analysis of the Vietnam War* (Novato, CA: Presidio Press, 1982), 1.

3. Department of the Army, FM 100-5, *Operations* (1 July 1976), i.

4. As related to the author by Lieutenant General (Ret.) John Cushman, then Commandant. The author was a student in the class of 1974-1975.

5. See comments of General (retired) Donn A. Starry in interview with John L. Romjue, TRADOC History Office, dated 19 March, 1993, 7-8. As late as the fall of 1972 there was no target for end strength or final force structure.

6. The best biographical account is an oral history, William E. DePuy, *Changing An Army: An Oral History of General William E. DePuy, USA, Retired,* Lieutenant Colonel Romie L. Brownlee and Lieutenant Colonel William J. Mullen III, eds. (Carlisle Barracks, PA: United States Military History Institute, n.d.). The interviews took place in 1979. See also *Selected Papers of General William E. DePuy*, compiled by Colonel Richard Swain (Fort Leavenworth, KS: Combat Studies Institute, 1994).

7. "Elimination of the Field Army," *Military Review*, LIII no. 10 (October 1973), 31-34.

8. General Starry used this formulation in a speech at Fort Leavenworth.

9. DePuy, *Changing an Army*, 90-102.

10. Omar N. Bradley, *A Soldier's Story* (New York: Popular Library, 1951), 298.

11. DePuy, *Changing an Army*, 202.

12. Ibid, 183.

13. "Implications of the Middle East War on U.S. Army Tactics, Doctrine and Systems," in *Selected Papers of General William E. DePuy*, 75-111.

14. Slide 12 in ibid., 85, and FM 100-5 (July 1976), 2-6.

15. Reprinted in *Selected Papers of General William E. DePuy,* 121. Copies of the letter are available in the William E. DePuy Papers, Box: Personal Files 1974-1975. Folder O [Ott]. U.S. Military History Institute, Carlisle Barracks, PA, and the DePuy collection in the History Office of Training and Doctrine Command, Fort Monroe, Virginia.

16. *Selected Papers of General William E. DePuy*, 122-36.

17. Ibid., 121. See DePuy's own interpretation in "FM 100-5 Revisited," in ibid., 303-9 (reprinted from *Army* 30, no. 11 [November 1980], 12-17).

18. General DePuy's reply to General Alexander Haig, dated 13 October 1975, in U.S. Army Command and General Staff College, Combined Arms Research Library (hereinafter CARL) Archive, Box 102A, folder CGSC 82, DTAC-010, "Al Haig's Comments on 1976 100-5." Haig had been DePuy's subordinate in the First Infantry Division.

19. Memorandum from Colonel Edwin G. Scribner, titled DOCTRINE DEVELOPMENT BY TRADOC; May 1973-December 1979, in CARL Archive, Box 106A, folder CGSC 82, DTAC-038, "Scribner Paper on Doctrinal Process."

20. Major Paul H. Herbert, *Deciding What Has to Be Done: General William E. DePuy and the 1976 Edition of FM 100-5, Operations*, Leavenworth Paper Number 17 (Fort Leavenworth, KS: Combat Studies Institute, 1988). Herbert is scrupulously fair to all parties in the dispute between the Command and General Staff College and its new

higher headquarters, Training and Doctrine Command.

21. The "A. P. Hill Draft" of FM 100-5, the Leavenworth attempt that did not receive approval of the TRADOC commander, is available for inspection in the Combined Arms Research Library Collection of superseded doctrinal publications.

22. Scribner memorandum, "Doctrine Development by TRADOC," 4-5.

23. Ibid., 2-3.

24. FM 100-5 (July 1976), ch. 4. On *mission orders*, see, in particular, 3-2.

25. Ibid., 5-2, 5-7.

26. Letter, General E. C. Meyer to General Donn A. Starry, 21 March 1981, in CARL Archive, Box 104A, folder CGSC 82, DTAC-026/022, "General Shy Meyer's Comments on Feb 1981 DRAFT of FM 100-5."

27. General Gorman's view of the DePuy years in *The Secret of Future Victories*, IDA Paper P-2653 (Alexandria, VA: Institute for Defense Analysis, 1992, reprinted by Command and General Staff College Press, Fort Leavenworth, KS, in 1994).

28. Starry Interview, 19 March 1993, 24-25.

29. Colonel Nicholas A. Andreacchio, "The Active Defense: A Snapshot," *Military Review* LXI, no. 6 (June 1981), 49-57.

30. Major Theodore T. Sendak, "The Airborne Antiarmor Defense," *Military Review* LIX, no. 9 (September 1979), 43-51.

31. Messages to Lieutenant General Donn A. Starry from General DePuy, Subject: Progress Report And Other Matters, and Concepts And Plans, in *Selected Papers of General William E. DePuy*, 185-91.

32. See General Donn A. Starry, "A Tactical Evolution—FM 100-5," *Military Review* LVIII, no. 8 (August 1978), 2-11.

33. Major Wesley K. Clark, "Winning the First Battle: Another Look at New Tactical Doctrine," unpublished essay by Staff College Class of 1975 first order of merit graduate. In CARL Archive, Box 105A, folder CGSC 82, DTAC-029, "Wes Clark Paper on Tactical Doctrine." Clark, also a Rhodes scholar, and first man in his class at West Point, is perhaps not exactly a representative observer.

34. Major Paul E. Cate, "Large Unit Operational Doctrine," *Military Review* LVIII, no. 12 (December 1978), 40-47; Starry, "A Tactical Evolution."

35. Interview with Brigadier General Donald R. Morelli, 12 January 1983, in CARL Archive, Box 106A, folder CGSC 82, DTAC-057, "BG Donald Morelli Interview," 15-16. The interview was done by Mr. John Romjue of the TRADOC History Office. See also, MSG 041900Z SEP 81, Subject: Institutional Stupidity, from General Starry to General Glenn K. Otis. Starry's message expresses outrage at a *Military Review* article written by an ADC to a Fort Leavenworth general and suggests, "The time for debate is over—those who don't want to sign up can just get out. . . ." In CARL Archive, Box 107A, folder CGSC 82, DTAC-074, "Anthony Coroalles Controversy."

36. William S. Lind, "Some Doctrinal Questions for the United States Army," *Military Review* LVII, no. 3 (March 1977) 54-65.

37. See the discussion of the army's initial response to Lind's essay by John Patrick, "Banned at Fort Monroe, or the Article the Army Doesn't Want You to Read," *Armed Forces Journal International* 114, no. 2 (October 1976), 26. Other discussion of the manual is on 4 and 23-4, by F. Clifton Berry, Jr., the editor; by Philip A. Karber, a prominent defense critic, "Dynamic Doctrine for Dynamic Defense," 28-29; and the Training and Doctrine Command response, 27-28.

38. William S. Lind, "Doctrinal Questions," 58.

39. Ibid., 64.

40. Colonel Wayne A. Downing, "Firepower Attrition Maneuver; U.S. Army Operations Doctrine: A Challenge for the 1980s and Beyond," *Military Review* LXI no. 1 (January, 1981), 64-73.

41. Colonel Robert E. Wagner, "Active Defense and All That," *Military Review* LX, no. 8 (August 1980), 4-13.

42. Richard Hart Sinnreich, "Tactical Doctrine or Dogma?" *Army* 29, no. 9 (September 1979), 16-19.

43. Ibid., 19.

44. Letters, *Army* vol. 29, no. 11 (November 1979), 4.

45. DePuy, "FM 100-5 Revisited."

46. Ibid., 308.

47. Note from Major General Richardson to Lieutenant General Meyer, dated 8 June 1979, in CARL Archives Box 102A, folder CGSC 82, DTAC-025/001, "1979 Letter Meyer to Starry on Problems in 1976 FM 100-5."

48. Letter from General E. C. Meyer to General Donn Starry, dated 13 June 1979, CARL Archive, Box 102A, folder CGSC 82, DTAC-007, "General Meyer on 1976 FM 100-5."

49. Ibid.

50. Ibid.

51. Ibid.

51. Letter from General Starry to General Meyer dated 26 June 1979. CARL Archive, Box 102A, folder CGSC 82, DTAC-024, "Individual Comments on 1981 DRAFT."

53. Ibid.

54. Ibid.

55. Interview with Lieutent Colonel (U.S.A. Ret.) Richmond B. Henriques, dated 14 May 1984, a Memorandum for Record by John Romjue, in CARL Archive, Box 106A, folder CGSC 82, DTAC-055, "LTC Richmond Henriques Interview."

56. General Donn A. Starry, "Extending the Battlefield," *Military Review* LXI, no. 3 (March 1981), 31-50.

57. FM 100-5 (July 1976), ch. 8. Morelli interview, 12 January 1983, 9.

58. As vice president of Ford Aerospace.

59. Roger J. Spiller, "War History and the History Wars: Establishing the Combat Studies Institute," *The Public Historian* 10, no. 4 (Fall 1988), 65-81.

60. Message, Cdr TRADOC FT Monroe VA//atrn-m//, For AIG 7573, Subject: The Writing of Doctrine, DTG 292030 APR 80. Message, SSO TRADOC to SSO LVN-WORTH, From Gen Starry, CG TRADOC, FT Monroe VA, For LTG Richardson, DCG TRADOC, FT LVNWORTH, Subject: The Writing of Doctrinal And Training Liturature, DTG 252300Z JUN 81. CARL, Archive, BOX 105A, folder CGSC 82, DTAC-027, "TRADOC Doctrinal Process."

61. Interview with Colonel Wass de Czege on the Development of AirLand Battle Doctrine, 16-17 April 1984, dated 24 April 1984 and Interview with Lieutenant Colonel L. D. Holder on AirLand Battle Doctrine, dated 8 May 1984, in CARL Archive, Box 106A, folder CGSC 82, DTAC-041, "Starry Letters—Various." Both interviews are memoranda for record by John Romjue. See also Huba Wass de Czege, "Army Doctrinal Reform," ch. 7 of Asa A. Clark IV, Peter W. Chiarelli, Jeffrey S. McKitrick, and James

W. Reed, eds., *The Defense Reform Debate* (Baltimore: Johns Hopkins University Press, 1984), 101-20.

62. Department of the Army, FM 100-5, *Operations* (20 August 1982), 2-1 to 2-2.

63. See Wass de Czege interview, 24 April 1984, and Lieutenant Colonel (P) Huba Wass de Czege and Lieutenant Colonel L. D. Holder, "The New FM 100-5," *Military Review* LXII, no. 7 (July 1982), 53-70.

64. FM 100-5 (August 1982), 2-1.

65. FM 100-5 (August 1982), 7-13 and 11-4.

66. Lieutenant General William R. Richardson, "Winning on the Extended Battlefield," *Army* 31, no. 6 (June 1981), 35-42. In a letter written to General Starry on 22 July 1981 Richardson reported, "The college is fully committed to working on the AirLand Battle, with respect to both the Integrated Battlefield (and its full accomplishment within tactical instruction) and the Extended Battlefield and how all elements must synchronize to defeat an enemy throughout the full depth of that battlefield." Copy of letter in possession of author.

67. See Holder Memo dated 13 October 1981, TAB B to a Staff Action, Subject: Cdr TRADOC's Directed Changes to FM 100-5 dtd 29 Oct 81, in CARL Archive, Box 105A. Action makes clear that what General Otis was concerned with was changing the army's view of the concept of maneuver.

68. See for example, Colonel Wallace P. Franz, "Operational Concepts," *Military Review* LXIV, no. 7 (July 1984), 2-15.

69. Colonel Arthur F. Lykke, Jr., ed., *Military Strategy: Theory and Application* (Carlisle Barracks, PA: U.S. Army War College, 1981).

70. Memorandum drafted by Lietenant Colonel Holder, Subject: Inclusion of the Operational Level of War in FM 100-5, dated 14 August 1981, in CARL Archive, Box 106A, folder CGSC 82, DTAC-041, "Starry Letters Various." See also Morelli interview, 21.

71. FM 100-5 (August 1982), 2-3.

72. All quotations from ibid.

73. Described in U.S. Army Combined Arms Center 1982-83-84, *Annual Historical Review* (Fort Leavenworth, KS: CAC History Office, 1989).

74. James J. Schneider, *The Structure of Strategic Revolution: Total War and the Roots of the Soviet Warfare State* (Novato, CA: Presidio, 1994), *passim*.

75. Source unknown. The quotation is still used in the School of Advanced Military Studies Command briefing.

76. Edward N. Luttwak, "The Operational Level of War," *International Security,* 5 no. 3 (Winter 1980/81), 61-79. A copy of the Net Assessment Paper is in CARL Archive, Box 102A, folder DTAC-018 "Luttwak/Canby on Operational Art."

77. Luttwak, "Operational Level of War," 61.

78. Ibid., 79.

79. V. Ye. Savkin, *The Basic Principles of Operational Art and Tactics*, translated and published by the U.S. Government Printing Office under the auspices of the United States Air Force in the series *Soviet Military Thought,* no. 4 (Washington, DC, 1982). Dr. Jim Schneider has pointed out to me that this trend began with the translation of Marshal Sokolovsky's *Strategy* in the 1960s.

80. Summers, *On Strategy*.

81. Carl von Clausewitz, *On War*, trans. by Michael Howard and Peter Paret, (Princeton, NJ, 1976), 128.

82. Michael Howard, "The Forgotten Dimensions of Strategy," *Foreign Affairs* 57, no. 5 (Summer 1979), 975-86.

83. CARL Archive Box 107A, folder CGSC 82, DTAC-070/001, "TRADOC Briefing to General Otis on Levels of War, Mar. 1982." A cover note is dated 23 March 1982. For War College thoughts at the time see, Colonel Wallace P. Franz, "Raising Art of War to Higher Level of Operative Excellence," *Army Times*, 42nd year, no. 26 (8 February 1982), 21.

84. Slide titled "Operational Level," in "TRADOC Briefing on Levels of War."

85. Slide titled "Reasons for Developing Levels of War Concept," in ibid.

86. Slide titled "Strategic Level," in ibid.

87. That is largely the point of Summers's book, *On Strategy*.

88. Wass de Czege informed the author that he received a copy of Luttwak's essay from General Richardson. See also, Morelli interview, 12 January 1983, 15-6.

89. See two papers by Sinnreich in the CARL Archive: Memorandum Thru Director Of Requirements, For Deputy Chief Of Staff For Operations And Plans, Subject: AirLand Battle Issues—Information Memorandum, dated 26 April 1983, in Box 113A, folder CGSC 83, SAMS-012, "AirLand Battle Implementation"; and Memorandum Thru Deputy Commandant For Commandant, Subject: "Marketing AirLand Battle Doctrine in NATO," dated 28 December 1984, in Box 102A, folder CGSC 82-DTAC-012, "Extended Battlefield."

90. See for example, Alex A. Vardamis, "German-American Military Fissures," *Foreign Policy* no. 34 (Spring 1979), 87-106.

91. U.S. Senate, 95th Congress, 1st Session. Committee Print. *NATO and the New Soviet Threat*. Report of Senator Sam Nunn and Senator Dewey F. Bartlett to the Committee on Armed Services, United States Senate, January 24, 1977. S 202-1.

92. General Starry's comments on the difficulties dealing with the differences between the German leaders' public and private judgments in Starry interview, 19 March 1993, 38-39.

93. Ibid. See also General Bernard Rogers, supreme Allied commander Europe, "NATO Strategy: Time to Change?" *The Alliance Papers No. 9*. Published by the Atlantic Council of the United States in cooperation with the United States Mission to NATO.

94. Memorandum For Reviewers Of FM 100-5, Subject: The Nature and Reasons for Changes in This Edition, dated 1 July 1985. CARL Archive, Box 124A, folder CGSC 86, SAMS-017, "Wass de Czege Paper on Reasons for Revising FM 100-5, 1 Oct 1985." (July Memo is an enclosure.)

95. Ibid., 2 (paragraph 4).

96. See Lieutent Colonel L.D. Holder, "A New Day for Operational Art; May Be Most Important Doctrine Change Since World War II," *Army* 35, no. 3 (March 1985), 22-32.

97. Glantz, a Soviet expert, brought together at these seminars senior veterans of the war on the Eastern Front who addressed the conduct of operations in light of contemporary defense problems.

98. See Professor James J. Schneider, "The Legacy of V. K. Triandafillov," in introduction to V. K. Triandafillov, *The Nature of the Operations of Modern Armies*, trans. by William A. Burhans, edited by Jacob W. Kipp (Newbury Park, Ilford, U.K.: Frank Cass, 1994), xxv-xxvi.

99. See, for example, Richardson's instruction to Wass de Czege that the final draft was to be hand-carried to General Rogers by the TRADOC chief of staff, Major General Penzler, in Memorandum For Record, Subject: FM 100-5 IPR with General Richardson on 24 September 1985, dated 30 September 1985, in CARL Archive, Box 124A, folder CGSC 86, SAMS-017, "Wass de Czege Paper on Reasons for Revising FM 100-5, 1 Oct 1985," 2 (paragraph 4).

100. U.S. Army. Field Manual FM 100-5, *Operations* (May 1986), i.

101. Ibid., 9-11.

102. Ibid., 10.

103. Ibid.

104. This is the thrust of Raymond Aron's interpretation of *On War*, accounting thereby for the revision begun in 1827. Raymond Aron, *Clausewitz, Philosopher of War* (London: Routledge and Kegan Paul, 1983), especially, "Preface" viii, and ch. 3, "The Final Synthesis and the Strategic Debate," 61-70.

105. FM 100-5 (May 1986), 181-82.

106. Summers, *On Strategy*, 1.

107. C. P. Snow, *The Light and the Dark. I. Strangers and Brother*, Omnibus Edition (New York: Charles Scribners' Sons, 1972), 244. I owe this citation to Dr. Roger Spiller, George C. Marshall Professor of History at the Army Command and General Staff College.

108. See my discussion of this problem in the Gulf War in Richard M. Swain, *"Lucky War", Third Army in DESERT STORM* (Fort Leavenworth, KS: Command and General Staff College Press, 1995), 182-83.

9

"The Revolution in Military Affairs": Its Implications for Doctrine and Force Development Within the U.S. Army[1]

Stéphane Lefebvre, Michel Fortmann, Thierry Gongora

Discontinuity and permanent, fundamental, and rapid change characterize a revolution.[2] To be successful, a revolution requires adaptation from the organizations and people it affects as well as the latter's recognition and legitimation of change. Of course, "the idea *itself* of a revolution creates new conditions, including threats to existing structures (and bureaucracies)."[3] The concept of a new revolution in military affairs (RMA) has yet to be fully developed and widely accepted, notwithstanding bold, forward-looking thinking and concrete initiatives from prominent military leaders.[4] The problem is that some analysts and policy-makers are drawn to the past to understand change or cling too ardently to immediate issues. They assume that change will be linear or that these issues are signposts of the future, when they should instead look at what produced these new trends and seek conclusive evidence. We are thus at a juncture of continuity and change.[5] The jury is still out, awaiting consensus among its members before propelling what many believe to be a RMA forward.

It is interesting to note here that the Soviet military was first thinking of a new RMA as early as the late 1970s. The sharp pen of Marshal Nikolai Ogarkov,[6] whose examination of the impact of emerging technologies led him to reassess the nature of future warfare, raised eyebrows among many. They were unable to see that the empirical evidence was only the outcome of a revolution, and not its precursor. Also dubbed military technological revolution (MTR), Ogarkov's RMA focused on the interaction between technology and the nature of warfare, while the current RMA, at least from an American point of view, is broader in scope.[7] Now that the superpower competition no longer exists, the motivation to achieve technological supremacy is winding down just as the "means of doing so accelerates."[8]

To comprehend the implications of a new RMA and to maximize its potential, it is necessary to understand first its purpose and nature. Knowing its character and core elements will assist policy-makers in implementing and exploiting such a revolution.[9] A RMA may thus serve several purposes. These may be internal (such as restructuring the Department of Defence) or external (such as attaining strategic objectives). A RMA may also be seen simply as a process to adapt to continuing changes or a means to filter new technologies. Beyond these purposes, a RMA can only be relevant if it addresses both the present and future security risks facing the nation.

If a RMA serves several purposes, it means that there probably are various types of RMA. Jeffrey Cooper lists three types.[10] The first is driven by military, scientific, or technological discoveries and developments.[11] It is the most popular understanding of the current RMA, probably as a result of the formidable display of technologies during the Gulf War, a display which in fact seems to have been more evolutionary than revolutionary.[12] Few doubt, however, that the enabling element of the RMA has been the current information revolution along with major advances in sensor technologies, avionics, guidance, stealth technology, electronic warfare systems, biological sciences,[13] and simulated training. The second type is problem-solving driven and is reflected in profound operational and organizational innovation that aims at solving a specific strategic problem. The German blitzkrieg strategy was of that type.

Finally, the third type is of a holistic nature and results from fundamental economic, social, and political transformations. A military revolution is not solely a military phenomenon; it is also shaped by broader social and political changes.[14] James Rosenau believes that such fundamental transformations are currently taking place. Indeed, he qualifies the end of the Cold War, the Gulf War, the dissolution of the Soviet Union, and other similar developments as outcomes of that type of revolution. Beneath these visible manifestations of change, he identifies underpinnings that will no doubt influence the future of war. These foundations include a shift in the sources of authority from traditional to performance criteria of legitimacy, the weakening of the state, the amazing increase in personal skills, the proliferation of subnational organizations and groups, and a shift from hard-core security concerns to economic concerns.[15] The impact of such transformations on the military institution is clear to Rosenau:

> It follows that any organization founded on clear lines of command and unquestioned compliance with orders is presently being undermined. . . . [S]ocieties [are] less reliant upon and less respectful of their military institutions even as the very complexity that undermines their reliance leads them to seek out their militaries to perform new tasks that are also a consequence of greater complexity and extensive change . . . [t]he military have moved from behind closed doors into the public arena, acting not as agents of the state but as claimants on its resources. . . . [T]he politization . . . of armed forces does seem likely to grow as their roles diminish and as the war weariness of publics deepens . . . military personnel are no longer silent and obedient with respect to their own welfare. . . . The conventional lines of authority do not follow the distribution of expertise among the ranks.[16]

A 1993 Center for Strategic and International Studies (CSIS) study offered a definition which combined the first two types of RMA: "A *Revolution in*

Military Affairs represents a fundamental advance in technology, doctrine or organization that renders existing methods of conducting warfare obsolete."[17] In other words, fundamental changes in doctrine and organization, to which are added enabling capabilities (such as information dominance) and executing capabilities (such as smart weaponry) make a revolution. Alvin and Heidi Toffler are close in their definition to the third type identified by Cooper. They believe that we are now in the "Third Wave," the postindustrial and knowledge wave of economic and social development which follows the Agrarian and Industrial waves.[18] However, the third wave coexists with the two previous ones, which means that today's wars may be waged in all of these.[19] Some observers reinforce that point by asserting that there is even a return to first wave warfare. They cite as evidence the increasing number of failed states and transnational violence, like that produced by drug cartels. Martin van Creveld supports that assessment when he writes that the weakening of the state system will lead to low-intensity conflicts in which irrelevancy will characterize advanced military technology.[20] One can think here of the difficulty in conducting high-tech intelligence gathering against an Iraq that "went stoneage" in its nuclear weapons programs. "Information-based warfare," Martin Libicki writes, "works best against industrial-based warfare and much less well against pre-industrial warfare."[21] He immediately adds, however, that in the near future urban warfare is likely to be a much more important preoccupation than unconventional rural conflict. Indeed, demographic changes in the Third World point to increased urbanization and use of urban dwellings by insurgent and other paramilitary groups.[22] Finally, the noted historian Sir Michael Howard cogently outlined the danger inherent in a conflict between first and third wave armies:

> Western societies have learned how to kill on an enormous scale, but they may still fight at a disadvantage against agrarian armies who have not forgotten how to die and know well enough how to kill. The Vietnam War and the recent experience in Somalia indicate that if those agrarian age armies are well-led, and if their leaders develop superior strategies, they can still prevail.[23]

This new emphasis on low-intensity conflict or conflict short of war noted above and which would oppose first and/or second to third wave armies may potentially limit the extent and scope of the RMA. However, as conflicts short of war are also transforming themselves from popular liberation wars to peace enforcement in failed states, so, too, are emerging new forms of ideological and commercial insurgency from quasi-political "gray area phenomena" such as narco-terrorism.[24] There is thus an argument to be made in favor of a greater role for new information technologies. Because opponents in conflicts short of war will likely have as one of their major objectives to change the minds of "enemy" policy-makers and the general population, advanced information technologies will be sought to deny them information and to compete with them on the same grounds but with better capabilities. In noncombatant evacuation operations, for example, new information technologies could be used to help locate and evacuate noncombatants, and unmanned aerial vehicles (UAVs) could conduct evacuation route reconnaissance. In relief operations, new helicopter technology (such as the V-22 Osprey) could, for example, increase the ability of relief agencies to reach refugees and resupply endangered civilians in remote areas.[25] Finally, nonlethal weapons currently in development could be used to reduce civilian casualties where and when force is necessary.[26] On the other hand, as Metz and Kievit suggest, technology specifically designed for conflict short of war is unlikely, at least in the short to middle term, to receive full or even significant funding. The prospect of being involved in that type of conflict may only result in the U.S. Army disengaging from these troubled spots.[27] This is especially so if belligerents discover and capitalize on the fact that the U.S. public's sensitivity to casualties is the American Achilles' heel.[28]

As the enabling element of the RMA, information technologies will provide the warrior with "unlimited amounts of information acquisition, processing, storage, and transmission capability" in packages that are small and inexpensive.[29] Information technologies will, at the operational level, be used to "synchronize integrated operations conducted at high-tempo, with high lethality and high mobility, throughout the depth and extent of the theater."[30] Simultaneity of actions will blur the distinctions between the strategic, operational, and tactical levels and alter what has heretofore been known as battle command. Coherence will need to be maintained across space and time. Operational effectiveness will therefore be contingent on widely distributed intelligence and on appropriate operational concepts and organizational structures.[31]

The development of information-distributed systems will be fundamental. Only these can put intelligence where it is needed in real-time. Decentralization will make these systems more robust against accidental failure, offer a much smaller signature than larger systems, and greatly complicate enemy countermeasures.[32] These systems—available in the coming 20 years in the form of sensors, emitters, microrobots, small reconnaissance platforms, intelligent munitions and microprojectiles, and so on—will number in the millions. In concert, with the interconnectivity of a mesh, they will be able to detect, track, and target with a high degree of accuracy. Destroying such a mesh would be arduous, as it will have no center of gravity. Libicki even argues the mesh thus created threatens the future of large and heavy system platforms, and the organizations that acquire and support them.[33]

Department of Defence (DOD) planners often have been accused of being too narrowly focused on heavy formations and the threat from regional powers in their efforts to revitalize the military's roles and missions in the post-Cold War era. With only a limited number of planning contingencies, the capability to project power, in whatever form, will become a paramount concern.[34]

What were heretofore strategic capabilities can now be used at all three levels of war, thus increasing space and diminishing time. This implies greater complexity, a shorter decision time for operational commanders, initiative at lower levels of command, decentralization of combat decision-making, and an emptying of the battlefield as it expands in all dimensions. The fact that technology has enhanced maneuverability and ranges has transformed the linear nature of the battlefield into a nonlinear or disengaged one.[35] In order to fight effectively in a nonlinear environment, synergy/jointness will have to be fully achieved, and dispersion, high lethality, rapid tempo, and leader initiative will be required. Finally, terrain retention will no longer be the primary objective; rather, the characteristic of nonlinear combat will put greater emphasis on enemy destruction.[36]

U.S. ARMY DOCTRINE AND THE RMA

Given the nature of the current RMA, what principles should guide doctrine formulation? Michael J. Mazarr proposes four: information dominance, synergy, disengagement, civilianization. Each is evolutionary but, when they are combined, their effect is revolutionary.[37] The context within which these principles will find themselves applied cannot be ascertained, however. Because of the chaotic nature of world politics, its many interdependent actors, and other variables, Mazarr asserts, "the United States cannot make any decisive assumptions about the precise enemies or conflicts it will face during the next two decades."[38] Thus, not only doctrine but also force structure and procurement will be affected by such a view of world politics. It follows that the general principle of keeping and sustaining forces and capabilities at every level of warfare is a very important one. What this means is that instead of looking for particular enemies to prepare against, the U.S. Army should prepare to face enemies at any of three levels of technological expertise: high technology, hybrid, or low technology.[39]

What will shape the RMA is the necessity for the U.S. Army "to modify and create technologies and force structures within the overarching doctrinal framework that add to warfighting effectiveness, while enhancing, or at the very least not diminishing, OOTW (Operations Other Than War) capabilities."[40] Development of a coherent OOTW war and noncombat operations (NCO) doctrine has until recently been conducted in an uncoordinated fashion, more as an ad hoc response to contingencies than as a deliberate expression of the army's forward-looking thinking.[41] Recent capstone doctrinal documents have been integrating OOTW and NCO alongside more war-fighting missions.[42] A presidential decision directive (PDD) on peace operations and a new army field manual on peacekeeping have been issued in 1994, and a new national military strategy incorporating for the first time peacekeeping and noncombat operations came out in March 1995. Another major document, Joint Publication 3-07, *Joint Doctrine for Military Operations Other Than War*, will be published by mid-1995. Notwithstanding the DOD inspector general's assessment that military training for peace operations calls for special training, the RMA is likely to provide the army with more flexibility in dealing with the political or psychological environments encountered in OOTW and NCO. The branches of service that may benefit the most from the information revolution are obviously civil affairs and psychological operations (PSYOPS), whose members are found predominantly in the reserve and guard components. This fact, however, may limit the force structure's adaptability for peace operations.[43]

Since doctrine development is fundamental for the exploitation of a RMA, one would expect guidance on this activity from the top down. The Office of the Secretary of Defence (OSD) has been accused of lacking a vision statement and of not providing overarching guidance on future concepts.[44] Without such vision and guidance, developing sound doctrine, and then developing the force, are much more difficult endeavors. As Andrew Krepinevich observed,

> In the absence of clear strategic guidance from the White House, the Pentagon bureaucracy has proceeded as most bureaucracies do when left on their own in a very difficult operating environment: they have attempted to "fit" the new situation to existing planning and resource allocation processes. The result is a defence program that is oriented on the most *familiar* threats, as opposed to the *greatest* or *most likely* threats.[45]

The development of the Bush administration's base force was typical of such an exercise, with General Colin Powell, chairman of the joint chiefs of staff, determined to leave his imprint on the reshaping of the military in a new international environment.[46] Powell's strategic vision encompassed scenarios that went beyond the diminished Soviet threat to regional and contingency responses to non-Soviet threats. However, he did not eliminate all the rigidity in DOD's strategic, program, and operations planning for large-scale contingencies. DOD's deliberate planning today is flawed because it only has a few detailed plans and no clear testing of procedures for adaptive planning for ad hoc contingencies. The standard scenarios do not really confirm the existence of holes, thus leading budget-cutters to authorize spending only for what is barely needed to meet the main regional contingencies envisaged in the Bottom-Up Review (BUR). The quick production of plans to meet a wide range of political-military objectives, however, faces at least two major obstacles:

> One obstacle to planning for prompt adaptiveness is the U.S. military's understandable antipathy toward ad hoc operations undertaken without careful consideration of potential consequences and development of both political consensus and determination. Still another obstacle is the military's desire to go into any conflict with overwhelming force so that casualties can be minimized and objectives achieved decisively.[47]

Again, it seems that uncertainty is best dealt with by focusing planning on the range of capabilities that the U.S. Army could encounter in entire regions or types of scenario. Adaptive planning certainly improved under Powell's chairmanship, but there is a need for more practice, experimentation, and learning to address nonstandard scenarios that might occur.[48]

A Canadian liaison officer at the U.S. Army Training and Doctrine Command (TRADOC) recently commented that much of the army doctrine community has an intellectual and futurist orientation.[49] The chief of staff of the army, General

Gordon R. Sullivan, has followed the example of General Donn A. Starry, TRADOC commander between 1977 and 1981, in instilling such an orientation. Starry was the one who first dispatched the army toward the third wave "by acquiring the latest communications technologies and precision weaponry, and by the forcing of decision-making . . . to lower levels in the organization."[50] Following Starry's belief that knowledge would become the military "capital" of the future, Sullivan has been a strong proponent of moving the army into the information age.

Sullivan characterizes the industrial age army as one preoccupied by standardization,specialization,professionalization,synchronization,concentration, maximization, and centralization. The aim of that type of army was the destruction of the enemy and its support base. The information age, in contrast, is defined by less hierarchical learning organizations, with the network as a model and knowledge as "capital." It is primarily preoccupied by nearly simultaneous, continuous, and short-run production. Its products are mass-customized, precisely targeted, and near instantaneously distributed. Agreeing with many analysts, Sullivan recognizes that the information age, or the third wave, will coexist with vestiges of the industrial and agrarian periods, thus increasing complexity and uncertainty in international relations.[51] In the information age, Sullivan asserts, common perception of the battlefield or shared situational awareness will be possible and essential to conduct operations, "resulting in the near-simultaneous paralysis and destruction of enemy forces, war-making capability, and information networks through the depth of a theater."[52]

It was only in February 1995, following the lead of the army (!), that the Pentagon engaged itself fully—though without a specific deadline—in defining its vision of information warfare, a vision that should guide the development and incorporation of new technologies into the force structure. Led by the director J6 (Command, Control, Communications and Computers), the Joint Staff has established two teams, one of experts from defence laboratories and research centers and another of military experts, to identify and determine how to use new information technologies.[53] Preceding the tangible efforts of the J6, however, was the formation of a Revolution in Military Affairs Steering Group under the collective leadership of Admiral William Owens, the vice chairman of the joint chiefs of staff; Paul Kaminski, the acquisition chief at the Pentagon; and Walter Slocombe, the principal undersecretary of defence for policy. It was announced on 1 March 1995 that within a month the Steering Group would receive recommendations on how to encourage innovative thinking and on the correlation of strategy and tactics with technological advances.[54]

The exploitation of information technologies through synergy/jointness is fundamental. Since the Goldwater-Nichols Department of Defence Reorganization Act of 1986, joint doctrine has not progressed far enough to satisfy jointness proponents. What joint doctrine actually needs is a capstone document similar to the Army's Field Manual 100-5, *Operations*, or to the air force Manual 1-1, *Basic Aerospace Doctrine of the United States Air Force*.[55]

Field Manual (FM) 100-5, *Operations*, the U.S. Army capstone doctrine document, has been subject to profound and important changes since its 1976 iteration, which basically confined U.S. forces to a costly attrition model of warfare. Firepower was the name of the game, to the detriment of maneuver. These inadequacies were corrected by the next iteration, in 1982, which introduced the concept of AirLand Battle, a "winning" doctrine which combined doctrine and technology. The 1986 edition of FM 100-5 officially introduced the operational level of war between the strategic and tactical levels into doctrine. Finally, the latest edition, promulgated in the summer of 1993, gives doctrine development the leading role in driving technological developments.[56] It is expected that the next edition of FM 100-5 will fully capture the characteristics of the information age.[57]

To prepare the ground for this next edition, to begin immediately experimenting with new technologies and force structures, and to guide the army toward the future, TRADOC was required to prepare a document encapsulating current and future changes liable to affect the battlefield. The result was TRADOC Pamphlet 525-5, *Force XXI: A Concept for the Evolution of Full-Dimensional Operations for the Strategic Army of the Early Twenty-First Century*, which is today's concept of the future. Divided into four chapters ("The Challenge of the Future"; "The Future Strategic Environment"; "Future Land Operations"; and "Implications on Doctrine, Training, Leader Development, Organizations, Material, and Soldiers"), it endeavors to explain "how flexible, agile, rapidly tailored units will be increasingly possible because of our ability to share and move timely information among quality soldiers, leaders and units."[58]

U.S. ARMY FORCE DEVELOPMENT AND THE RMA

Force development "comprises the design of organizations and material to accomplish combat and peacetime tasks. It is the function that creates the military potential needed to carry out the tasks."[59] The first serious post-Cold War effort at force development was Bush's base force, which was rapidly superseded by the Clinton administration's BUR, directed by Secretary of Defence Les Aspin. The 1993 BUR offered few elements suggesting awareness of a RMA. But since the BUR's focus was on the coming seven years rather than the coming two decades, this is excusable. The fact is that serious thought is being given to the issue and major initiatives are being implemented throughout DOD.[60] The BUR's scenarios were likewise limited, offering very few elements for future force development.[61] It identified the dangers to the United States as being major regional threats and nuclear proliferation, thus virtually eliminating the possibility of major-power war.[62] From these dangers, it derived three general missions: the capability to wage major regional conflicts (MRCs) in Iraq and South Korea, the ability to conduct peacekeeping or peace enforcement operations, and the capacity to maintain a forward presence where vital interests could be at stake.

The major conceptual problem is that the type of regional adversary used in the BUR appears to possess the same Iraqi doctrine and force structure that existed during the Gulf War. It is likely that these forces would look rather different even five years in the future.[63] Moreover, given the likely continuing military superiority of the United States, "America's adversaries will have great incentives to adopt a very unconventional approach."[64] That is, they have learned that limited U.S. objectives can be countered by means other than force, and, if force were to be used, it will not likely be on U.S. terms. This dimension was not fully explored in the BUR. The BUR is thus not optimized for the future but for a projection of a nonevolutionary set of circumstances found in the recent past.[65] In any case, force development planning can hardly be effective in the information age if it relies on threat analyses rather than on capabilities necessary for uncertain future environments. Put simply, one should look at what probable opponents could do since one cannot know what they will do.[66] That line of reasoning was followed by Sullivan in his view that the army should be smaller but more capable, equipped with modern technology, well trained and led, and subjected to up-to-date doctrine with an organization within which technology and doctrine are compatible.[67] He believes that the information age will force leaders to think very differently about how they go about things. For instance, the amount of available information will be colossal and made available over shorter periods of time. Leaders at all levels will have to make decisions faster and execute them "over greater distances and in decreasing time."[68]

To wage the information war (jamming, blinding, deceiving, and destroying emitters, sensors, and other nodes as well as affecting the enemy's key information systems such as air control, financial networks, and space-based communications), Martin Libicki proposes the creation of a separate information corps that would comprise the J6 and computers, the Defence Information Systems Agency (DISA), the Defence Material Agency, the United States Space Command (USSPACECOM), the various military intelligence agencies, and the controlling organizations of platforms such as JSTARS and others strategically important sensors. This information corps would promote jointness in a critical field.[69] It would develop and exploit an integrated image of battle space, devoting itself to data integration standardization from the start, rather than allowing the different services to try to merge information collection and dissemination systems.[70] A similar proposal was echoed recently by Colonel Charles Miller, head of the Air Force Staff's Strategic Planning Division: "Maybe this [a distinct specialty group for information warfare] will be the Special Operations Forces of the future, or maybe there will be a distinct information [commander in chief]."[71]

The basic combat unit is envisaged by Mazarr to be the reinforced brigade.[72] Because of the applicability of Sun Tzu's tenets,[73] Mazarr also believes that special operations forces may be the organizational model of the future:

> Special operations units are small, agile, flexible, able to take on a wide range of missions, highly trained and motivated, and imbued with the need for decentralized initiative. They use stealth and guile rather than brute force to achieve their objectives.

These same principles will dominate the doctrines of the regular U.S. military in the years to come.[74]

Military officers also echoed the concern that the current basic combat structure centered around the division was archaic and hardly relevant for the future types of war in which the army is likely to be involved. Their proposal is to have smaller, flexible, and easily deployed formations.[75] In that light, some have emphasized the need for more light armored units as they will likely be first to deploy in future contingency operations.[76]

Leadership, skills development, and collective training stand out as especially important for most analysts. As Mazarr observes, "Only highly intelligent, superbly trained, well-equipped troops with high morale and wide experience will be able to flourish in the incredibly demanding atmosphere of future war."[77] This is all the more important as war is ultimately "an affair of the heart" where such factors as courage, honor, duty, loyalty, and fear have nothing "to do with technology, whether primitive or sophisticated."[78] Sullivan largely shares this assessment. With Colonel James Dubik, he wrote that "only the highest quality soldiers, leaders, staffs, and organizations who understand the importance of speed and precision in information processing and applications will be able to succeed" in the information age.[79] However, he added, "Even in the information age, war will remain a human endeavour, subject to emotion and characterized by the shedding of blood and the effects of chance."[80]

CURRENT U.S. ARMY INITIATIVES

To shape its army of the future and maintain its core competencies (to fight in joint/combined environments, to be versatile, to operate across the continuum of military operations, to be strategically deployable, to be expandable, to be able to win rapidly and decisively), the Army initiated in March 1994 a major force redesign: Force XXI. Based on TRADOC's Pamphlet 525-5, Force XXI, the effort has three sequential and simultaneous axes: digitization, joint venture, and table of distribution and allowances (TDA)/institutional army (IA). To support the Digitization axis and leverage information age technology, the army has created a Digitization Office. The main effort, joint venture, whose objective is to ascertain what the future force will look like, is a collaborative effort between the U.S. Army TRADOC, the Army Material Command (AMC), United States Forces Command (USFORCESCOM), the U.S. Army Intelligence and Security Command, the U.S. Army Information Systems Command, the United States Army Medical Command, and the army staff. It must attain fielding decisions by fiscal year 2000, design and validate the results of advanced warfighting experiments (AWEs), and keep the Army abreast of the implications of digitized full-dimensional operations.[81] Led by the army deputy chief of staff for operations and plans, the TDA/IA axis represents the effort to develop a total army plan including reservists and civilians.[82] The three axes are supported by two processes: the battle laboratories (battle labs), and the Louisiana maneuvers

(LAMs), which try the concepts the labs develop. Exercising maximal leadership, the LAMs and battle labs are the means by which the army will rapidly incorporate change into its force structure and core competencies.[83]

There are five battle labs, each assigned specific battle dynamic concepts that represent the areas most affected by the third wave: mounted/dismounted battle space, lethal and survivable early entry, depth and simultaneous attack in all three dimensions, battle command, and responsive and versatile combat service support.[84] The battle space lab represents the recognition by the U.S. Army that the future battlefield will increasingly be nonlinear. Battle space is "that volume determined by the maximum capabilities of a unit to acquire and engage the enemy." Contrary to the former area of influence that characterized that situation, the new battle space "is not defined by time, boundaries, graphics, countermeasures or other physical constraints."[85] Because of the expected extension of the battlefield in space, time, and purpose, and the elimination of the close, deep, and rear operations concepts,[86] the battle space lab concentrates on reassessing the relationship between fire and maneuver.

The battle command lab focuses on what the commander must be able to do throughout the full cycle of his mission. To be successful in the information age, the commander must be able to visualize his present and future battle space with the support of a fully digitized and lean battle command support team. Ultimately,

> high-tech sensors will be employed to see the enemy in all conditions, day or night. Sensor data will be fused, processed, correlated and compressed to create information which can then be distributed using digital communications technology and appropriately presented by command and control systems to effect decisions.[87]

There will, thus, be a shift from the current command post fixations, large tactical staffs, and process-oriented control.[88] Moreover, "individual soldiers will be empowered for independent action because of enhanced situational awareness, digital communications and a common view of what needs to be done."[89] The responsive and versatile combat support battle lab supports the Army Material Command (AMC) strategic infrastructure, which focuses on "three core competencies—logistics power projection, technology generation and application, and acquisition excellence."[90] The AMC's third wave development and acquisition processes are guided by six principles, the most important of which is focussing on capabilities rather than equipment. They are expected to accommodate new and maturing technologies rather than focus on an end item defined much earlier. They must also use industry standards whenever possible,[91] bring systems into alignment with the open architecture, encourage industry to use army standards wherever the army is leading, and take a holistic approach.[92]

The LAM, the joint venture's synchronization mechanism, represents the test bed of the future digitized force. It is modeled after the Louisiana maneuvers used to prepare the U.S. Army for the Second World War. EXFOR (Experimental Force-First Brigade, Second Armored Division), the first digitized formation, should be fielded in 1996.[93] A digitized division should follow in 1997, and a corps in 1999. The first advanced warfighting experiment (AWE) that will lead to such a force was conducted in April 1994 and consisted of the digital connection of 144 systems.[94] Four AWEs will be conducted in 1995. AWE Focused Dispatch will concentrate on mounted force operations. AWE Warrior Focus will compare the performance of a conventional light task force with a fully digitized dismounted force. AWE Prairie Warrior will focus on the Mobile Strike Force (MSF)—a simulated experimental-size division—in a nonlinear environment. The intent of the MSF is "to build a land combat force from the battle labs' input that uses the organization, material and operational concepts derived" in TRADOC Pamphlet 525-5.[95] Finally, AWE Theater Missile Defence will test operations against mobile missile launchers, command and control nodes, and missile support equipment. In addition, Field Artillery will continue its development of a Battlefield Coordination Element (BCE), which will interface the air component command with the army during joint operations.[96]

Critics assert that Force XXI is a hypothetical project, overly dependent on future technologies that the army will unlikely be able to afford. In addition, Force XXI results, reflecting the passage from hierarchical command and rigid battle space to internetted command and fluid battle space, will have to overcome the conservative nature of the army.[97] For Andrew Krepinevich, director of the Defence Budget Project, the bulk of the army "seems to be much more concerned with making existing Army organizations more effective than they are, than in creating entirely new kinds of organizations and doctrine to exploit this explosion in technology."[98]

The extension of the battlefield into a nonlinear one, in which distinctions between deep, close, and rear battles are absent, is a subject of intense debate between the army and the air force, and one of the more tedious to resolve by Congress's Commission on Roles and Missions of the Armed Forces.[99] Criticizing the other services for not understanding the "revolution in modern air warfare," air force officers would prefer to keep the battlefield bounded, with the Joint Forces Air Component commander (JFACC) in charge of authorizing strikes by any weapon systems beyond a 50- to 100-kilometer boundary.[100] Such an option would undermine Army programs such as the Army Tactical Missile System (ATACMS) and the Comanche helicopter. The army position on this issue is that "one commander, who's focused on the objective in an integrated battlefield must have the ability to orchestrate all elements of combat power to win as decisively as possible with minimum loss to the force."[101] In other words, the army advocates a joint service perspective on the issue of hitting the enemy in his depth.[102]

CONCLUSION

Even if the RMA is at hand, its nature and implications are more open to debate. One cannot doubt that technology is making giant strides and that its possibilities are far-reaching. The push for technological progress that leads into the third wave, however, is also partly an answer to fundamental political, economic, and social changes. Facing reduced budgets, a smaller pool of recruits, and uncertainty on the international scene, the military has few alternatives but to do much more with a lot less. Information and other advanced technologies thus present themselves as saviors.

The United States has taken a tremendous lead in the development of information warfare technologies. But as the commercial availability of related or similar products increases over the years (via extension or duplication), the U.S. military will have to keep pushing forward the boundaries of its technological knowledge and expertise by taking full advantage of dual-use technologies.[103] DOD has already recognized that it can no longer afford defence-unique technologies. As a result, the DOD fiscal year 1996-1997 budget calls for increased reliance on dual-use commercial technologies, products, and processes. Among the initiatives being pursued are a restructuring of the defence acquisition system, the integration whenever possible of commercial technologies into new systems, and an increased investment in research and development of dual-use technologies.[104]

The future direction to be given to the Pentagon's basic research efforts was outlined in February 1995 when the director of defence research and engineering released the report *Defence Basic Research: A Prospectus*. Information systems are given a high priority, along with virtual reality, artificial intelligence, and smart materials.[105] The 1994 *Army Modernization Plan* update follows the same direction, but because of underfunding, upgrading rather than introducing new systems is likely to be forced upon the service.[106] In a recent interview with *Defence News*, Togo West, the army secretary, commented that "[f]or the kind of program we need to equip the Army, [the research, development and acquisition budget] doesn't even come close. It isn't even holding [our] own. It is continuing to spiral down. What we have done is to try to hold the best we can."[107]

At present, the army has serious deficiencies in the areas of data distribution, common user systems, command and control on the move, computer automation, and distributive imagery intelligence.[108] With the current initiatives, however, the army is positioning itself to meet its chief of staff's five strategic modernization objectives: to win the information war, dominate maneuvers, execute precision strikes, protect the force, and project and sustain combat manpower.[109] In fact, the army presently leads the other services in these areas. It was only in February 1995, for instance, that the U.S. Marine Corps launched its major effort to define its future in relation to the RM. Project Vision 21, the nickname given to that effort, will assess the impact of developments in technology, economics, psychology, and society on the service.[110] The U.S. Navy formed its first panel—the Chief of Naval Operations' Executive Panel—to study naval warfare innovations in 1994. A final report was due in May 1995. Indications are that a greater use by the navy of advanced munitions and battlefield surveillance systems will be recommended.[111]

A consensus among military officials is slowly emerging that doctrine and force structure must reflect the probable technological capabilities of various types of opponents rather than their intentions or actual force posture. The technological capabilities that can be found in each of the Tofflers' waves thus offer a better rationale than planning on the basis of who is going to threaten the United States in an unspecified number of years.[112] Robert Jervis's warning against forecasting is apposite:

> Forecasting is quite difficult in the absence of a solid theoretical base [in international relations]. . . . [I]t is unlikely that a single variable will determine the course of events in world politics and even if it has in the past, it may not continue to do so in the future. . . . [F]amiliarity with social sciences findings can influence the way actors behave. Decision-makers may even learn not to behave as they have in the past thereby diluting the predictability of generalizations based on earlier behaviour. . . . [T]here will always be ample room for the values, preferences, beliefs, and choices of decision-makers to play some role. To the extent that foreign policy is strongly influenced by these individual level factors, predicting their future values is a highly dubious undertaking. . . . Contingencies [such as change and accidents] matter and since we do not know what they will be, prediction is exceedingly difficult.[113]

Uncertainty in a trisected world, a world in which states of the first, second, and third waves coexist, implies that planning must be adaptive to any future contingencies and not centered around only two or three foreseen scenarios reminiscent of past conflicts.

The impact of the revolution will be profound on battle command. It will take years before the new battle command processes brought to bear by the information age will be fully understood, let alone developed.[114] But as John Guilmartin, Jr., concluded in his recent study on technology and war, one should be reminded that "whatever the technology, war remains as Carl von Clausewitz characterized it, a test of will and faith. Do not lose sight of that reality."[115]

NOTES

1. The views expressed in this paper are those of the authors and do not in any way reflect the official position of their respective organizations.

2. Steven Metz and James Kievit, *The Revolution in Military Affairs and Conflict Short of War* (Carlisle Barracks, PA, July 1994), 1.

3. Jeffrey R. Cooper, *Another View of the Revolution in Military Affairs* (Carlisle Barracks, PA, July 1994), 21-22.

4. "[W]e must develop a *comprehensive general theory of military revolutions* set within the context of broader notion of global politics and security. Currently, there is no accepted definition of RMAs or even agreement on which historical transformations constituted revolutions." Metz and Kievit, *Revolution in Military Affairs*, 28.

5. David Jablonsky, *The Owl of Minerva Flies at Twilight: Doctrinal Change and Continuity and the Revolution in Military Affairs* (Carlisle Barracks, PA, May 1994), 2, notes that with change there is always continuity, given "a tendency of people to repeat and continue their way of doing things as long as possible." On immediate issues and signposts, see Paul Bracken, "Future Directions for the Army," in idem., *Whither the RMA: Two Perspectives on Tomorrow's Army* (Carlisle Barracks, PA, 1994), 1.

6. See Marshal Nikolai Ogarkov, *Istoriya uchit bditel'nosti* (History Teaches Vigilance) (Moscow, 1985), and *Vsegda v gotovnosti k zashchite* Otechestva (Always in Readiness to Defend the Fatherland) (Moscow, 1982).

7. Jablonsky, *Owl of Minerva*, 1. For current Russian views on future war, see General of the Army Mahmut Gareev, "The Art of War in a Changing World," *International Affairs* (Moscow) (June 1994), 75-83; and Mary C. FitzGerald, "The Russian Image of Future War," *Comparative Strategy*, 13 (1994), 167-80.

8. Martin Libicki, *The Mesh and The Net: Speculations on Armed Conflict in a Time of Free Silicon*, McNair Paper no. 28 (Washington, DC, March 1994), 6.

9. Cooper, *Revolution in Military Affairs*, 3-4.

10. Ibid., 20-21.

11. Libicki, *The Mesh and The Net*, 3, contends that "[o]ver time radical changes in technology are understood to involve radical changes in the organization of work and society as well." It is also interesting to note that military-technological innovation has shifted to the commercial from the government sector. Cf. Bracken, "Future Directions," 6.

12. John Guilmartin, Jr., "Technology and Strategy: What Are the Limits?" in Michael Howard and John Guilmartin, Jr., *Two Historians in Technology and War* (Carlisle Barracks, PA, July 1994), 29. Metz and Kievit, *Revolution in Military Affairs*, 1, wrote that "[t]he Gulf War was widely seen as a foretaste of RMA warfare, offering quick victory with limited casualties. As a result, most attention has been on the *opportunities* provided by RMA rather than its *risks, costs,* and *unintended side effects*."

13. "Today, two RMAs may be underway simultaneously. The first (and more mature) is electronic. . . . The second (and potentially more profound) RMA is biotechnological, including genetic engineering and advanced behaviour-altering drugs. . . . It may ultimately be the combination of the two that proves truly revolutionary." Metz and Kievit, *Revolution in Military Affairs*, 32.

14. Michael J. Mazarr, *The Revolution in Military Affairs: A Framework for Defense Planning* (Carlisle Barracks, PA, June 1994), 2.

15. Lieutenant Colonel (U.S.M.C.) T. X. Hammes, *The Evolution of War: A Fourth Generation* (Kingston, Ont., June 1994), 2-6, outlined basically the same factors.

16. James N. Rosenau, "Armed Force and Armed Forces in a Turbulent World," in James Burk, ed., *The Military in New Times: Adapting Armed Forces to a Turbulent World* (Boulder, CO, 1994), 25-61.

17. Quoted in Jablonsky, *Owl of Minerva*, 7.

18. The Industrial Wave's technological impact on the nature of war is widely recognized by historians. Sir Michael Howard, "How Much Can Technology Change Warfare?" in Howard and Guilmartin, *Two Historians*, 1. Howard notes: "During the agrarian age, the only fundamental changes that occurred in the conduct of war were the results of social and political factors rather than technological innovation."

19. Hammes, *The Evolution of War*, 27, notes that fourth generation warfare—war in the information age—will include tactics and techniques from earlier generations.

20. On these views see Jablonsky, *Owl of Minerva*, 7-11.

21. Libicki, *The Mesh and The Net*, 85.

22. Jennifer Morrison Taw and Bruce Hoffman, *The Urbanization of Insurgency: The Potential Challenge to U.S. Army Operations* (Santa Monica, CA, 1994), 15, observe that

> [e]ven if insurgents choose not to base their operations in urban areas, they can nonetheless take advantage of urbanization. Rural-based insurgencies are finding cities increasingly lucrative targets. Whereas cities were once the culmination of the revolution, the proliferation of urban areas—and the inability of governments to defend them all—has made cities relatively simple targets that can yield substantial political rewards for relatively little effort. Insurgent groups can disrupt energy and telecommunications facilities, draw international attention, demonstrate the inability of the government to protect its people, and recruit from among the disaffected population. Even those insurgencies that remain based in rural areas can take advantage of urbanization by increasing their reliance on terrorism against urban targets.

23. Howard, "Can Technology Change Warfare?" 8.

24. Metz and Kievit, *Revolution in Military Affairs*, 4.

25. Ruth Wedgwood, "The Smart New Weaponry Can Also Help Keep Peace," *International Herald Tribune*, 16 December 1994.

26. Rudimentary nonlethal weapons were deployed to Somalia in February 1995. See Pat Cooper, "U.S. Tests Nonlethal Weapon Policy in Somalia," *Defense News* 10 (27 February-5 March 1995), 28.

27. Metz and Kievit, *Revolution in Military Affairs*, 4-7, 12.

28. For an excellent study on this subject, see Benjamin C. Schwarz, *Casualties, Public Opinion and U.S. Military Intervention* (Santa Monica, CA, 1994).

29. Libicki, *The Mesh and The Net*, 7. E.R. Hooton has concluded likewise:

> The days when navies, or even companies, developed their own dedicated hardware, if not over, are clearly drawing to a close because civilian desire to exploit information technology means that a wide range of hardware such as microprocessors, display systems and local area networks is available at relatively low cost. For this reason navies and manufacturers are now beginning to exploit commercial off-the-shelf (COTS) technology.

See his preface to *Jane's Naval Weapon Systems*, Issue 15 (Coulsdon, U.K., June 1994), n.p. The Army Science Board not long ago recommended the use of Internet compliant protocols to exchange information on the digitized battlefield. See Otto Guenther and Robert F. Giordano, "Enabling Technologies and Advanced Concepts for the Digitized Force XXI," *Army Research, Development and Acquisition* (PB 70-94-6) (November-December 1994), 24.

30. Cooper, *Revolution in Military Affairs*, 30.

31. Ibid., 30, 33.

32. Libicki, *The Mesh and The Net*, 12-13, 32.

33. Ibid., 50.

34. Cooper, *Revolution in Military Affairs*, 10.

35. Jablonsky, *Owl of Minerva*, 29-36; Mazarr, *Revolution in Military Affairs*, 18-19. Since warfare is going to be nonlinear, Mazarr argues that the notion of concentration of mass will be replaced with one emphasizing concentration of fire.

36. Major J. Marc LeGare, *Paradigm Found—The Nuclear and Nonlinear Battlefields* (Fort Leavenworth, KS, February 1993), 16-33.

37. Civilianization is understood to reflect the larger emphasis now put on reserves and militia, the increasing use of nonlethal weapons to reduce casualties and collateral damages, and the importance of civilian rather than military information technologies.

Mazarr, *Revolution in Military Affairs*, 4, 23-25.

38. Ibid., 7-8.

39. See Major John W. Nicholson, Jr., *Blainey and the Bottom-Up Review: Increased Potential for Miscalculation and War in the 21st Century* (Fort Leavenworth, KS, May 1994), 3.

40. Jablonsky, *Owl of Minerva*, 39.

41. For example, "Both doctrine and training fail to sufficiently link intelligence operations, psychological operations, and civic action in urban counterinsurgency efforts, even though the effectiveness of each type of operation can be multiplied by coordination with the others." Taw and Hoffman, *Urbanization of Insurgency*, 15.

42. Jennifer Morrison Taw and Robert C. Leicht, *The New World Order and Army Doctrine: The Doctrinal Renaissance of Operations Short of War?* (Santa Monica, CA, 1992), 12-34.

43. Sarah B. Sewall (deputy assistant secretary of defense for peacekeeping and peace enforcement policy), "Peace Operations: A Department of Defense Perspective," *SAIS Review* XV (1995), 121.

44. Raoul Henri Alcala, "Guiding Principles for Revolution, Evolution, and Continuity in Military Affairs," in Bracken, *Whither the RMA*, 32.

45. Andrew F. Krepinevich, Jr., "The Clinton Defense Program: Assessing the Bottom-Up Review," *Strategic Review* XXII (1994), 23. Nicholson, *Blainey and the Bottom-Up Review*, 15, reflects similar negative feelings: "This budget driven decision-making process is the major domestic factor affecting the debate on military force structure and strategy."

46. "Rather than waiting for the President to enunciate a new national security strategy, which he and the Secretary would then be charged with implementing, General Powell believed that it was his responsibility to press for a change in strategy in response to the changes in the strategic environment. He also thought that, as Chairman, he should provide programming direction to the Services." Lorna S. Jaffe, *The Development of the Base Force 1989-1992* (Washington, DC, July 1993), 49.

47. Paul K. Davis, "Institutionalizing Planning for Adaptiveness," in Paul K. Davis, ed., *New Challenges for Defense Planning: Rethinking How Much Is Enough* (Santa Monica, CA, 1994), 79.

48. Ibid., 81, 89.

49. Lieutenant Colonel J. P. Sweetnam, "New Thinking in the U.S. Army: The Louisiana Manoeuvres, Battle Laboratories and the Third Wave Army," *Canadian Defence Quarterly* 24 (1994), 23.

50. Ibid., 24.

51. General Gordon R. Sullivan and Colonel James M. Dubik, *War in the Information Age* (Carlisle Barracks, PA, June 1994), 1-11.

52. Ibid., 13.

53. Pat Cooper and Robert Holzer, "Pentagon Rethinks Art of War," *Defense News* 10 (20-26 February 1995), 3, 28.

54. "Board Soon to Deliver RMA Recommendation," *Defense News*, 10 (6-12 March 1994), 2.

55. Alcala, "Guiding Principles," 27.

56. See Jablonsky's overview, *Owl of Minerva*, 25-27.

57. Sullivan and Dubik, *Information Age*, 17.

58. Brigadier General Morris J. Boyd and Major Michael Woodgerd, "Force XXI Operations," *Military Review* LXXXIV (November 1994), 18.

59. Alcala, "Guiding Principles," 32.

60. Mazarr, *Revolution in Military Affairs*, 3.

61. Alcala, "Guiding Principles," 18. See also Les Aspin, Secretary of Defense, *The Bottom-Up Review: Forces for a New Era* (Washington, DC, 1 September 1993), 5.

62. For William R. Thompson, "The Future of Transitional Warfare," in Burk, *Military in New Times*, 88, the demise of major-power war "within the next generation or two is less than a sure thing. Indeed, if we assume it cannot happen again we are more likely to facilitate its reoccurrence than if we assume that, unfortunately, it remains very much a possibility."

63. Krepinevich, "Clinton Defense Program," 19.

64. Ibid., 20.

65. "U.S. Military Strategy and Force Posture for the 21st Century," *RAND Research Brief* (September 1994).

66. See Bruce W. Bennett, Sam Gardiner, and Daniel B. Fox, "Not Merely Planning for the Last War," in Davis, *New Challenges*, 477-89.

67. Sullivan and Dubik, *Information Age*, 15. Congressman Ronald V. Dellums, former chairman of the House Armed Services Committee, expressed the same opinion recently: "If we focus too much on the dangers rather than the potential opportunity, then we will become paralyzed and fail to secure policies that can better generate stability and security for the United States." See Dellums, "Toward the Post-Transition World: New Strategies for a New Century," *SAIS Review* XV (Winter-Spring 1995), 94.

68. Sullivan and Dubik, *Information Age*, 19.

69. Mazarr, *Revolution in Military Affairs*, 12, defines "closeness" as "the ability of different services, branches, and weapons to fight effectively together, to marshal their unique capabilities into a whole that is greater than the sum of its parts."

70. Libicki, *The Mesh and The Net*, 51-57, 62-63. On page 69, he adds,

> As wars are currently fought, the need for a data corps is, while perhaps inevitable, not necessarily urgent. . . . The logical conclusion, nevertheless, is that DOD should make steps to form an Information Corps. The argument is that a corps would promote jointness where it is critically needed (information interoperability), elevate information as an element of war, develop an information warrior ethos and curriculum, and heighten DOD attention to the global civilian net.

71. Cooper and Holzer, "Pentagon Rethinks Art of War," 28.

72. Mazarr, *Revolution in Military Affairs*, 22.

73. For instance, "that the acme of success is to avoid engagements and impose one's will through skilful maneuver and exploitation of weaknesses." Lieutenant Colonel David Todd, "Gird for Information War," *Defense News* 10 (6-12 March 1995), 20.

74. Mazarr, *Revolution in Military Affairs*, 31.

75. For example, Lieutenant Colonel Curtis L. Newcomb, *Principles of Future Army Force Structure Design* (Carlisle Barracks, PA, April 1993), 20.

76. "Light armor's strengths of speed, mobility, firepower, and agility enable it to seek decisive results in the enemy's rear and flank areas." Major Alan M. Mosher, *Light Armor in Deep Operational Maneuver: The New Excalibur* (Fort Leavenworth, KS, May 1994), 5.

77. Mazarr, *Revolution in Military Affairs*, 29.

78. Martin van Creveld, *Technology and War From 2000 B.C. to the Present* (New York, 1989), 314.

79. Sullivan and Dubik, *Information Age*, 13.

80. Ibid., 15.

81. Boyd and Woodgerd, "Force XXI Operations," 17-18.

82. Thomas G. Conway, "America's Army. . . Into the 21st Century. Explaining the Army Chief of Staff's Message," *Army Research, Development and Acquisition* (January-February 1995), 11-13.

83. As the commanding general of the U.S. Army Material Command noted, "Force XXI is neither a specific organization nor a particular warfighting doctrine. Rather, it is a process of experimentation and exploration of emerging technologies." See General Leon E. Solomon, "Shaping the U.S. Army Material Command for Force XXI," *Army Research, Development and Acquisition* (January-February 1995), 2.

84. Sweetnam, "New Thinking," 26. See also Lieutenant Colonel John R. Brooks and Captain John M. Fahey, "Battle Focus Center (IEW Integration)," *Military Intelligence* (October-December 1993), 4-8; Captain Vincent J. Colwell, "Battle Command Battle Lab—Fort Gordon," ibid., 9-10; and Captain David Hiles, "Depth and Simultaneous Attack Battle Lab—Fort Sill," ibid., 11-4.

85. Boyd and Woodgerd, "Force XXI Operations," 22.

86. They will be fused into a single, seamless battle space. Brigadier General Leo J. Baxter, "Field Artillery Vision 2020," *Field Artillery* (December 1994), 12. See also LeGare, *Paradigm Found*, 21-22.

87. Guenther and Giordano, "Enabling Technologies," 21.

88. Major James C. Madigan and Major George E. Dodge, "Battle Command: A Force XXI Imperative," *Military Review* LXXIV (November 1994), 30.

89. Boyd and Woodgerd, "Force XXI Operations," 21.

90. Solomon, "U.S. Army Material Command," 3.

91. The Army Research Laboratory (ARL) "will take cutting-edge private-sector technologies and apply them to the unique military environment in four areas: Sensing. . . . [d]istribution. . . . [a]nalysis. . . . [a]ssimilation." James R. Predham, "Army Research Laboratory Contribution to Force XXI," *Army Research, Development and Acquisition* (January-February 1995), 16.

92. General Gordon R. Sullivan, "Force XXI: Digitizing the Battlefield," *Army Research, Development and Acquisition* (November-December 1994), 2-3.

93. Interconnected digitized systems will include the Ground Based Sensor, the Forward Area Air Defense Command and Control System, the OH-58D Kiowa Warrior helicopter, the Paladin Howitzer, the AH-64D Apache Longbow attack helicopter, the improved Single Channel Ground-to-Air Radio Systems and prototype Battlefield Combat Identification Systems, the Hunter medium-range unmanned aerial vehicle, advanced communications systems from the battle labs, and M1A1 with applied software and hardware compatible with those of the other platforms. See Jason Glashow, "U.S. Army Places EXFOR on Top of Weapon Priority List," *Defense News* 10 (6-12 March 6-12, 1995), 14.

94. Sullivan, "Digitizing the Battlefield," 3.

95. Lieutenant Colonel James K. Greer, "Experimenting with the Army of the 21st Century," *Field Artillery* (December 1994), 43.

96. Major General John A. Dubia, chief of field artillery, "Force XXI and the Field Artillery: State of the Branch 1994," *Field Artillery* (December 1994), 2-4.

97. Daniel G. Dupont and Richard Lardner, "Force XXI: The Long and Winding Road to the Army of the Future," *Armed Forces Journal International* 132 (October 1994), 45-46.

98. Quoted in Dupont and Lardner, "Force XXI," 46.

99. Congress of the United States, Congressional Budget Office, *Reducing the Deficit: Spending and Revenue Options* (Washington, DC, February 1995), 15:

> The Congress, in the National Defense Authorization Act for Fiscal Year 1994, established a Commission on Roles and Missions of the Armed Forces. The commission's charter is to review all aspects of the organization of the Department of Defense for possible efficiencies and improvements. It will review such matters as duplication among the services in performing military missions, as well as the consolidation of support activities such as training, maintenance, and intelligence gathering. The commission's report is due in May 1995.

100. Jason Glashow and Robert Holzer, "AF Role Proposal Riles Other Service Leaders," *Defense News* 10 (27 February-5 March, 1995), 6.

101. Interview with General John H. Tilelli, Jr., vice chief of staff of the army, in P. S. Hollis, "The Army and FA Challenges of Designing Force XXI," *Field Artillery* (December 1994), 8.

102. Jason Glashow and Robert Holzer, "U.S. Army Battles USAF Turf Grab", *Defense News*, 10 (6-12 March 1995), 10.

103. Libicki, *The Mesh and The Net*, 120-21. This was a role that DARPA, the Defense Advanced Research Projects Agency, once took upon itself, amid much debate. See William J. Broad, "Pentagon Wizards of Technology Eye Wide Civilian Role," *New York Times* (22 October 1991), B5.

104. Department of Defense, press release on the FY 1996-1997 budget.

105. "Pentagon Reveals Research Priorities," *Defense News* 10 (27 February-5 March 1995), 2.

106. Robert Ropelewski, "Army Shifts Modernization Effort From Embedded to Upgraded Approach", *Signal* 49 (November 1994), 43-45.

107. Interview with Togo West in Jason Glashow, "Army Secretary Links Readiness, Technology," *Defense News* 10 (6-12 March 1995), 28.

108. Ropelewski, "Army Shifts Modernization Effort," 44-45.

109. Statement by General Gordon R. Sullivan, chief of staff, U.S., Before the Committee on Armed Services, U.S. House of Representatives, First Session, 103rd Congress, Fiscal Year 1994 Budget Proposals and the Posture of the United States Army, 31 March 1993, 19-22.

110. Robert Holzer, "Marine Generals Plot 21st Century Corps," *Defense News* 10 (13-19 February 1995), 30.

111. Robert Holzer, "U.S. Navy Study Promotes Precision Munitions," *Defense News* 10 (27 February-5 March 1995), 30.

112. Several senior officers reached that conclusion. For example, Major General John A. Dubia wrote that "[t]he pace of technological developments has made the 'sufficiency' method of designing the force [capabilities based on known threats] obsolete." See Dubia, "Force XXI," 1.

113. Thompson, "Transitional Warfare," 80. Jervis's arguments can be found in Robert Jervis, "The Future of World Politics: Will It Resemble the Past?" *International Security* 16 (1991/92), 39-73.

114. Boyd and Woodgerd, "Force XXI Operations," 27.

115. Guilmartin, "Technology and Strategy," 40.

10

Commentary on the Operational Art[1]

Charles F. Brower IV

"War, like Gaul," David Jablonsky reminds us, "is divided into three parts"—military strategy, operational art and tactics.[2] Occupying its central position along the continuum of warfare, operational art began to receive close attention from military theorists and practitioners only after the Napoleonic Wars of the nineteenth century changed the nature of warfare. That early attention had a decidedly Prussian and German accent until after the First World War, when Soviet theorists—in part reflecting German judgments on the topic—added their intellectual energies to the task of raising the operational art to the level of a distinct category of military theory. Anglo-Saxon appreciation of the operational level of war did not truly emerge until a quarter century after the Second World War. The resulting and remarkable American renaissance of doctrinal creativity in the 1970s and 1980s was stimulated by the loss of the war in Vietnam, influenced by the Soviet doctrinal example, and largely adopted by British allies. Today, contemporary military thinkers are musing, in the aftermath of the Cold War and victory in the Gulf, about the implications for operational art of what some see as a "Revolution in Military Affairs."[3]

This volume, based on the Royal Military College of Canada's Twenty-first Military History Symposium, explores and analyzes the operational art from an interesting variety of perspectives and methods of inquiry, though the approach is principally a historical one. In Chapter 2 John English traces insightfully the definitional and intellectual foundations of operational art from its Prussian roots to the present. His story of the sharpening understanding of the operational art remains focused on its role as the pivot between strategy and tactics.

Three chapters provide a set of diverse operational case studies for consideration. First, Bradley J. Meyer carefully analyzes Helmuth von Moltke's campaign plan for the 1870 Franco-Prussian War in terms of campaign objectives, operational goals, and the mechanics of moving and deploying large bodies of troops. William McAndrew, in Chapter 5, offers a critical assessment of the Canadian way of war in the Second World War, deftly exploring operations in the Italian campaign and in northwest Europe. Finally, Sabine Marie Decup interprets French operational methods from 1945 to 1970 with special attention to the wars in Indochina and Algeria.

A second trio of historians explores Soviet and American contributions to the operational art. Examining the Red Army's first decade, Jacob Kipp discusses in detail how Soviet doctrine evolved from emerging Soviet assessments of Marshal Mikhail Tukhachevsky's abortive 1920 Warsaw campaign during the Russian Civil War, while David M. Glantz establishes the intellectual context for the evolution of Soviet operational art from the Russian Revolution to the 1990s. In Chapter 8, Richard M. Swain provides a crisply written insider's assessment of the U.S. Army's doctrinal revolution in the 1970s and 1980s and the process by which it adopted and embraced the concept of the operational art. He shows that the U.S. Army's systematic study of Soviet military theory and practice informed its understanding of the operational art and influenced its final shape, an insight that Glantz also demonstrates to have worked reciprocally in the evolution of Soviet doctrine.

Chapter 9, the combined work of political scientists Stéphane Lefebvre, Michel Fortmann, and Thierry Gongora, investigates the implications for doctrine and force development in the American armed forces of the "revolution in military affairs" now thought ongoing by some. Strongly influenced by the arguments of those who see the prodigious strides in information technologies as fundamentally changing the conduct of warfare,[4] these authors lean in the direction of shaping future American doctrine and force structure on the basis of the technological capabilities of future adversaries rather than on a broader vision of the nature of future warfare.

This book presents not only a representative sample of the important national contributions to the development and practice of operational art, but also helps us to lift the veil shrouding what English calls "that gray area between tactics and strategy," to enrich our understanding of the how national perspectives influence the process of doctrinal evolution, and to maintain a balanced perspective on how nations and armies prepare themselves to adapt to revolutions in military affairs.

German flair in the operational art notwithstanding,[5] these authors in general rank the intellectual contribution of Soviet military theorists as more important to a clearer understanding of the operational art. Kipp, Glantz and English all note that Soviet General-Major A. A. Svechin was responsible in the 1920s for coining the term "operational art" and coming to grips with its essence, and Swain illustrates that the sophisticated and comprehensive body of Soviet military theory in existence in the 1970s both informed and stimulated American efforts to fill the operational void in its military doctrine. Svechin's metaphor for the concept was that of a bridge between tactics and strategy, with operational art acting as the link between a series of tactical successes to progress along the path provided by strategy. "Battle," he wrote in his 1926 classic *Strategy*, "is the means of the operation. Tactics are the material of operational art. The operation is the means of strategy, and the operational art is the material of strategy."

Better understanding of the essence of this relationship takes a large step toward avoiding the mistake of compartmentalizing warfare, a tendency that the preceding chapters suggest was all too common. Svechin's metaphor reminds us that the operational artist must be sensitive to the need to *link* strategy and tactics and to guard against preferential travel in only one direction. As Bradley J. Meyer shows, even as refined a practitioner of the operational art as Helmuth von Moltke may have been unduly preoccupied with the tactical-operational link to the detriment of the strategic-operational, a tendency that Gunther Rothenberg analyzed in greater detail in his classic essay in *Makers of Modern Strategy*.[6] And as John English points out, both Rothenberg and Dennis Showalter have shown that German operational thought continued this tendency to devolve downward toward the tactical and the operational levels well into the twentieth century, much to Germany's detriment strategically.[7]

Several authors in this volume illustrate the tendency toward the compartmentalization of warfare. McAndrew, for example, reveals that the Canadian way of war in the Second World War possessed the same tactical predisposition as that of the Germans, though not blessed with the same brilliance. He claims, correctly, that the Allied Italian campaign in 1943-1945 provides "a classic study of the disjunction between the operational and tactical levels of war."[8] Over time, the objective of the Italian campaign gradually came to be viewed by the combined chiefs of staff in terms of its ability to divert Nazi divisions from Normandy and the Eastern Front and to fix them in Italy; paradoxically, the operational goal to accomplish that object became driving the Germans out of Italy altogether. In a curious fashion, failure in the one became the requisite for success in the other.

Swain's account of the U.S. Army's development of Field Manual 100-5, *Operations*, characterizes the initial version published in 1976 as "a tactical manual, and a minor tactical manual at that." Focused at the battalion and brigade level, the camouflaged-covered 1976 version was a "how to fight" manual that provided few insights into the operational art or its relation to strategy. However, Swain finds that by 1986 a good deal had been learned about the operational art and its link to strategic goals, lessons that were incorporated into the third version of the manual. Indeed, some have worried that the new American operational outlook—greatly influenced by Colonel Harry Summers's critique of American strategy in the Vietnam War—had shifted too dramatically in the direction of the strategic-operational link. This trend may have been necessary corrective medicine for an army with a history of being fundamentally concerned with tactical questions; one also senses a warning to guard against the tendency for compartmentalization.

Swain's chapter and those of Kipp and Glantz also contribute to our understanding of how armies go about developing doctrine. Doctrine is not created in a vacuum but is affected by institutional culture and experience, Swain observes in a generalization that has relevance to both the American and Soviet cases. Unlike the Soviet Union, the United States, as a maritime and insular power, had little need for employing and maneuvering large armies for much of its national experience. Its maritime perspective imbued it with a strategic inclination, much as it did with Great Britain, and, in part, explains the late attention of the Anglo-Saxon powers to the operational level of warfare. John English reminds us that it was not until the United States found itself confronting large Soviet armies in Central Europe after the Second World War that the United States and its NATO allies began to appreciate and deal with the challenges of large unit operations embraced many decades earlier by Germans and Soviets.

As with strategic culture, national experience clearly played an important role in the evolution of operational art in the two cases. The loss of the Vietnam War traumatized the American military and stimulated intellectual and doctrinal soul-searching on the part of many American officers who pondered how an army undefeated on the battlefield could lose the war to North Vietnam.[9] Their conclusions led them both to a self-serving critique of policy which has not served the army well *and* to the realization that a means had to be found to link such a series of tactical successes to the larger strategic aim, and that link was operational art. More thoughtful American military theorists soon determined that the challenge in Vietnam was not, as Summers claimed, the absence of clearly defined political aims and guidance. Instead the challenge was to formulate suitable operational goals which would serve the political purpose.[10] This is no simple task—Meyer terms the derivation of operational goals "the most creative act of the operational commander"—but it is the key to excellence in the operational art. Swain shows that, fortunately for the evolution of American operational doctrine, the intellectual ferment in the American army in the aftermath of the Vietnam War channeled itself mostly away from blaming policy-makers and toward an introspective effort to understand and correct its own failures in the higher-level conduct of military operations in Vietnam.

In the Soviet case Kipp and Glantz both illustrate how Bolshevik experience in the Russian Civil War, conditioned by Marxist-Leninist ideology, meshed with lessons derived from Russia's experiences in the First World War to determine and sustain the emergence in the 1930s of Soviet concepts of "deep battle" and "the deep operation" which became the ideals of the Soviet operational art for the next half century. However, the road from interwar theory to Great Patriotic War practice proved a bumpy one. Arduous lessons learned from fighting the Nazis on the Eastern Front led, in the final phases of the war, to what Glantz characterizes as "a practical rebirth" of Soviet operational art as the Red Army rediscovered its interwar theory of the deep operation.

One is struck in these discussions of the evolution of the Soviet operational art by the relatively unintrusive role played by ideology. Kipp demonstrates that early Soviet assessments of the First World War and Civil War experiences were indeed colored by ideological romanticism of the Revolution being carried forward on the bayonets of the Red Army. That perspective was effectively counterbalanced, however, by the important contributions of "military specialists," *voenspetsy*, whose ideas carried weight precisely because of their knowledge of traditional military theory and practice learned in the service of the Tsar.[11] The result was a doctrinal evolution characterized by synthesis, abstraction, and balance, not revolutionary ideology.

Swain also endorses the American doctrinal process for its synthetic and abstract character, and applauds its imitative character. Indeed, it is difficult to underestimate the significance of the contribution that the intensive American study of Soviet doctrine in the 1970s and 1980s made to the flowering of the American interest in the operational art. That study was grounded in the American army's awareness of the importance of military history and military education, a commitment to the development of military theory resting upon the systematic study of past warfare that both Kipp and Glantz also identify as a fundamental elements of the Soviet doctrinal evolution.

The significance of the creative human dimension of the evolutionary process emerges clearly in the works of each of these authors. The intellectual energy and forceful advocacy of Depuy, Starry, Richardson, Wass de Czege, Holder, and Sinnreich, and of Tukhachevsky, Svechin, Varfolomeev, and Ogarkov permeate the pages and remind us that ideas and concepts take meaningful form and shape in specific ways not as a result of great, often unfathomable forces but as a result of human beings who are able to step forward and influence history's direction.

One final precautionary note emerges from this book. A cottage industry has emerged in the past few years debating the future of warfare and various aspects of what has come to be called the "revolution in military affairs." The way in which American and allied forces in the Gulf defeated Saddam Hussein's forces suggests to some that the revolution is now ongoing and has generated intense speculation about its future path. Fearful that the failure to adapt to a revolutionary change in the nature of warfare will dissipate America's preeminent position as the only global superpower, military analysts and soldiers alike have seized on the notion that the advanced technology of the information age has revolutionized warfare.[12]

Lefebvre, Fortmann and Gongora are not prepared to risk being unready for the challenges of third wave warfare and, accordingly, advocate forward-looking changes in American doctrine and force structure. Such an approach may, however, fall victim to the vice of another kind of compartmentalization. By assessing future warfare from too narrow a perspective, they risk minimizing the significance of other social, cultural, and political considerations which also shape the nature of war. It is unclear whether understanding American success in the Gulf War should be more relevant to our thinking about the nature of future war than is the American involvement in Somalia. A. J. Bacevich may very well be correct in warning that, in too ardently embracing technology as the means for harnessing future war, "soldiers are willfully blinding themselves to other powerful elements that shape warfare," with potentially disastrous consequences.[13]

NOTES

1. The views expressed in this chaper are those of the author and do not reflect the official policy or position of the Department of Defense or the U.S. government.

2. David Jablonsky, *The Operational Art of Warfare Across the Spectrum of Conflict* (Carlisle Barracks, PA, 1987), 5.

3. See, for example, Andrew J. Krepinevich, "Cavalry to Computer: The Pattern of Military Revolutions," *The National Interest* 37 (Fall 1994), 40-42; and David Jablonsky, "U.S. Military Doctrine and the Revolution in Military Affairs," *Parameters* 3 (Autumn 1994), 23-27.

4. General Gordon R. Sullivan and Colonel James M. Dubik, "War in the Information Age," *Military Review* LXXIV, no.4 (April 1994), 46-62.

5. Edward N. Luttwak, "The Operational Level of War," *International Security* 5, no.3 (Winter 1980/81), 61-103, esp. 67-73.

6. Gunther E. Rothenberg, "Moltke, Schlieffen, and the Doctrine of Strategic Envelopment" in Peter Paret, ed., *Makers of Modern Strategy From Machiavelli to the Nuclear Age* (Princeton, NJ, 1986), 296-325.

7. Dennis E. Showalter, "Total War for Limited Objectives: An Interpretation of German Grand Strategy," in Paul Kennedy, ed., *Grand Strategies in War and Peace* (New Haven, CT, 1991), 105-21.

8. Sabine Marie Decup finds a similar disjunction in French operational methods in Indochina and Algeria.

9. Harry G. Summers, Jr., *On War: The Vietnam War in Context* (Carlisle Barracks, PA, 1981), 1.

10. John M. Gates, "Vietnam: The Debate Goes On," *Parameters* 14 (Spring 1984), 15-24.

11. Condoleezza Rice, "The Making of Soviet Strategy" in Paret, *Makers of Modern Strategy*, 658.

12. See, for example, Alvin and Heidi Toffler, *War and Anti-War: Survival in the Dawn of the 21st Century* (Boston, 1993); Martin van Creveld, *The Transformation of War* (New York, 1991); David Jablonsky, *The Owl of Minerva Flies at Twilight: Doctrinal Change and Continuity and the Revolution in Military Affairs* (Carlisle Barracks, PA, 1994); and General Gordon R. Sullivan and Colonel James M. Dubik, *War in the Information Age* (Carlisle Barracks, PA, 1993).

13. A. J. Bacevich, "Preserving the Well-Bred Horse," *The National Interest* 37 (Fall 1994), 49.

Selected Bibliography

Alcala, Raoul Henri. "Guiding Principles for Revolution, Evolution, and Continuity in Military Affairs." in P. Bracken, ed., *Whither the RMA: Two Perspectives on Tomorrow's Army.* Carlisle Barracks, PA, 1994.

Alger, J. I. *The Quest For Victory: The History of the Principles of War.* Westport, CT, 1982.

Andreacchio, Colonel Nicholas A. "The Active Defense: A Snapshot." *Military Review* LXI, no. 6 (June 1981).

"L'Armée de l'Air en Algerie: la Lutte Anti-Guerilla." *Toute l'Aviation* 8 (1992).

Aron, Raymond. *Clausewitz: Philosopher of War.* London, 1983.

Aspin, Les. *The Bottom-Up Review: Forces for a New Era.* Washington, DC, 1 September 1993.

Bacevich, A. J. "Preserving the Well-Bred Horse." *The National Interest* 37 (Fall 1994), 49.

Battesti, M. "La Marine et la Guerre d'Indochine." *Revue Historique des Armées* 177 (1989).

Baxter, Brigadier General Leo J. "Field Artillery Vision 2020." *Field Artillery* (December 1994).

Bezkhrebty, M. I. "Sovmestnaia operatsiia—glavnaia forma sovremennykh boevykh deistvii." *Voennaia mysl'* no.7 (July 1979).

Biarnès, P. *Les Français en Afrique Noire de Richelieu à Mitterand.* Paris, 1987.

Bidwell, Shellford and Dominick Graham. *Fire-Power: British Army Weapons and Theories of War 1904-1945.* London, 1982.

Bigeard, M. *Pour une Parcelle de Gloire.* Paris, 1975.

Blumenson, Martin. "A Deaf Ear to Clausewitz: Allied Operational Objectives in World War II." *Parameters* 23 (1993).

"Board Soon to Deliver RMA Recommendation." *Defense News* 10 (6-12 March 1994).

Booth, Ken, ed. *New Thinking About Strategy and International Security.* London, 1991.

Boyd, Brigadier General Morris J. and Major Michael Woodgerd. "Force XXI Operations." *Military Review* LXXXIV, no.11 (November 1994).

Bozo, F. *La France et l'OTAN.* Paris, 1991.

Bracken, Paul. "Future Directions for the Army." In Paul Bracken, ed., *Whither the RMA: Two Perspectives on Tomorrow's Army.* Carlisle Barracks, PA, 1994.

Broad, William J. "Pentagon Wizards of Technology Eye Wide Civilian Role." *New York Times* (22 October 1991).

Brooks, Lieutenant Colonel John R. and Captain John M. Fahey. "Battle Focus Center (IEW Integration)." *Military Intelligence* (October-December 1993).

Brown, Ian. "Not Glamorous, but Effective: The Canadian Corps and the Set-piece Attack," *The Journal of Military History* 3 (July 1994).

Brownlee, Romie L. and William J. Mullen III, eds. *Changing an Army: An Oral History of General William E. Dupuy.* Carlisle Barracks, PA, 1988.

Carver, M. *War Since 1945*. London, 1990.

Cate, Major Paul E. "Large Unit Operational Doctrine." *Military Review* LVIII, no. 12 (December 1978).

Cherednichenko, M. "Nekotorye cherty sovremennogo voennogo iskusstva." *Voennaia mysl'*, no. 2 (February 1970).

Chervonabab, V. "Printsipy voennogo iskusstva i ikh razvitie" (Principles of military art and their development). *Voennaia mysl'*, no. 11 (November 1971).

Clausewitz, Carl von. *On War*. Edited and translated by Michael Howard and Peter Paret. Princeton, NJ, 1976.

Cohen, Eliot A. and John Gooch. *Military Misfortunes: The Anatomy of Failure in War*. New York, 1990.

Colwell, Captain Vincent J. "Battle Command Battle Lab—Fort Gordon." *Military Intelligence* (October-December 1993).

Congress of the United States, Congressional Budget Office. *Reducing the Deficit: Spending and Revenue Options*. Washington, DC, February 1995.

Cooper, Jeffrey R. *Another View of the Revolution in Military Affairs*. Carlisle Barracks, PA, July 1994.

Cooper, Pat. "U.S. Tests Nonlethal Weapon Policy in Somalia." *Defense News* 10 (27 February-5 March 1995).

Cooper, Pat and Robert Holzer. "Pentagon Rethinks Art of War." *Defense News*. 10 (20-26 February 1995).

Corum, James S. *The Roots of Blitzkrieg: Hans von Seeckt and German Military Reform*. Lawrence, KS, 1992.

Craig, Gordon A. *The Politics of the Prussian Army 1640-1945.* London, 1955.

———. *The Battle of Koeniggraetz: Prussia's Victory Over Austria, 1866.* New York, 1964.

Dabezies, P. "Les Interventions Françaises Outre-Mer." In D. David, ed., *La Politique de Défense de la France, Textes et Documents*. Paris, 1989.

Davis, Norman. *White Eagle, Red Star: The Polish-Soviet War, 1919-1920.* New York, 1972.

Davis, Paul K., ed. *New Challenges for Defense Planning: Rethinking How Much Is Enough*. Santa Monica, CA, 1994.

Dellums, A. "Toward the Post-Transition World: New Strategies for a New Century." *SAIS Review* XV (Winter-Spring 1995).

D'Este, Carlo. *Decision in Normandy.* London, 1983.

Devries, Lieutenant Colonel Paul T. "Manoeuvre and the Operational Level of War." *Military Review* LXIII, no. 2 (February, 1983).

Dick, Charles. *The Operational Employment of Soviet Armour in the Great Patriotic War*. Sandhurst, 1988.

Dick, C. J. "Soviet Doctrine, Equipment Design and Organization." *International Defence Review* 12 (1983).

Dietrich, Steve E. "The Professional Reading of General George S. Patton, Jr." *The Journal of Military History* 4 (October 1989).

Doise J. and M. Vaïsse. *Diplomatie et Outil Militaire 1871-1991*. Paris, 1992.

"Doktrina voennaia." *Sovetskaia istoricheskaia entsiklopediia* 3 (1976).

Donnelly, Christopher. "The Development of the Soviet Concept of Echeloning." *NATO Review* 6 (December 1984).

Downing, Colonel Wayne A. "Firepower Attrition Maneuver. U.S. Army Operations Doctrine: A Challenge for the 1980s and Beyond." *Military Review* LXI, no. 1 (January 1981).

Dubia, Major General John A. "Force XXI and the Field Artillery: State of the Branch 1994." *Field Artillery* (December 1994).

Duffy, Christopher. *Red Storm on the Reich.* New York, 1991.

Dupont, Daniel G. and Richard Lardner. "Force XXI: The Long and Winding Road too the Army of the Future." *Armed Forces Journal International* 132 (October 1994).

Dupuy, Colonel T. N. *A Genius for War: The German Army and the General Staff, 1807-1945.* London, 1977.

———. *The Evolution of Weapons and Warfare.* London, 1980.

Eddy, Colonel K. T. "The Canadian Forces and the Operational Level of War." *Canadian Defence Quarterly* 5 (April 1992).

Egorov, A. I. "Taktika i operativnoe iskusstvo RKKA na novom etape." *Voenno-istoricheskii zhurnal* no. 10 (October 1963).

Ellis, John. *Brute Force: Allied Strategy and Tactics in the Second World War.* London, 1990.

English, John A. "Central Army Group: Pillar of NATO Defence." *Janes 1982-83 Military Review.* London, 1982.

———. *The Canadian Army and the Normandy Campaign: A Study of Failure in High Command.* New York, 1991.

Erickson, John; Lynn Hansen; and William Schneider. *Soviet Ground Forces: An Operational Assessment.* Boulder, CO, 1986.

Falls, Cyril. *The Art of War From the Age of Napoleon to the Present Day.* Oxford, 1961.

Farago, Ladislas. *Patton: Ordeal and Triumph.* New York, 1975.

FitzGerald, Mary C. "The Russian Image of Future War." *Comparative Strategy* 13 (1994).

Fomin, N. N. "Sovetskoe operativnoe iskusstvo: Zarozhdenie i osnovnye etapy razvitiia." *Voennaia mysl'* No. 12 (December 1978).

Foures, A. *Au-Delà du Sanctuaire.* Paris, 1986.

———. "Caractères des Interventions Militaires Françaises Outre-Mer de 1960 à nos Jours." *Revue Historique des Armées* 169 (1987).

Franchini, P. *Les Guerres d'Indochine.* Vol. 1. Paris, 1988.

Franz, Wallace P. "Grand Tactics., *Military Review* LXI, no. 12 (December 1981).

———. "Raising Art of War to Higher Level of Operative Excellence." *Army Times* 42nd year, no. 26 (8 February 1982).

———. "Maneuver: The Dynamic Element of Combat." *Military Review* LXIII, no. 5 (May 1983).

———. "Operational Concepts." *Military Review* LXIV, no. 7 (July 1984).

Fuller, J. F. C. *The Foundations of the Science of War.* London, 1925.

———. *Decisive Battles: Their Influence Upon History and Civilization.* New York, 1940.

Gaget, R. *Commandos Parachutistes.* Paris, 1992.

Gareev, Mahmut. "The Art of War in a Changing World." *International Affairs* (Moscow)

(June 1994).

Garthoff, Raymond L. *Deterrence and the Revolution in Soviet Military Doctrine*. Washington, DC, 1990.

Gates, John M. "Vietnam: The Debate Goes On." *Parameters* 14 (Spring 1984).

Gaujac, P. *Histoire des Parachutistes Français*. Vol. 1. Paris, 1975.

Glantz, David M. "Soviet Operational Formation for Battle: A Perspective." *Military Review* 2 (February 1983).

———. "The Nature of Soviet Operational Art." *Parameters* 1 (Spring 1985).

———. *Soviet Tactics at Kursk, July 1943.* Fort Leavenworth, KS, 1986.

———. *The Soviet Conduct of Tactical Maneuver*. London, 1991.

Glashow, James. "Army Secretary Links Readiness, Technology." *DefenseNews* 10 (6-12 March 1995).

———. "U.S. Army Places EXFOR on Top of Weapon Priority List." *Defense News* 10 (6-12 March, 1995).

Glashow, James and Robert Holzer. "AF Role Proposal Riles Other Service Leaders." *Defense News* 10 (27 February-5 March, 1995).

———. "U.S. Army Battles USAF Turf Grab." *Defense News* 10 (6-12 March 1995).

Goerlitz, Walter. *History of the German General Staff.* Boulder, CO and London, 1985.

Golovchiner, B. "Nekotorye voprosy sovremennogo operativnogo iskusstva." *Voennaia mysl'* no. 10 (October 1961).

Golubev, A. "Obrashchena li byla v proshloe nasha voennaia teoriia v 20-e gody?" *Voenno-istoricheskii zhurnal* no. 10 (October 1965).

Grechko, A. A. "Voennaia nauka." *Sovetskaia voennaia entsiklopediia* 2 (1976).

Greer, Lieutenant Colonel James K. "Experimenting with the Army of the 21st Century." *Field Artillery* (December 1994).

Griffith, Paddy. *Battle Tactics of the Western Front: The British Army's Art of Attack 1916-18.* New Haven, CT, 1994.

Grosser Generalstab, Kriegsgeschichtliche Abteilung I. *Der deutsch-franzoesischeKrieg, 1870-1.* Erster Theil: Geschichte des Krieges bis zum Sturz des Kaiserreiches, Heft 1: Die Ereignisse im Monat Juli. Berlin, 1872.

———. *Moltke in der Vorbereitung und Durchfuehrung der Operationen.* Krieges geschichtliche Einzelschriften, Heft 36. Berlin, 1905.

Gudmundsson, Bruce I. *Stormtroop Tactics: Innovation in the German Army, 1914-1918.* New York, 1989.

Guenther, Otto and Robert F. Giordano. "Enabling Technologies and Advanced Concepts for the Digitized Force XXI." *Army Research, Development and Acquisition* (PB 70-94-6) (November-December 1994).

Guilmartin, Jr., John. "Technology and Strategy: What Are the Limits?" In M. Howard and John Guilmartin Jr. *Two Historians in Technology and War.* Carlisle Barracks, PA, July 1994.

Hammes, Lieutenant Colonel (USMC) T. X. *The Evolution of War: A Fourth Generation.* Kingston, Ontario, June 1994.

Handel, Michael I. ed. *Clausewitz and Modern Strategy*. London, 1989.

Harris, J. P. and F. H. Toase. *Armoured Warfare*. London, 1990.

Haycock, Ronald and Keith Neilson. *Men, Machines, and War*. Waterloo, Ontario, 1988.

Hebert, Major Paul H. *Deciding What Has to Be Done: General William E. DePuy and the 1976 Edition of FM 100-5, Operations*. Leavenworth Paper Number 17. Fort Leavenworth, KS, 1988.

"L'Hélicoptère Lourd dans les Opérations de Maintien de l'Ordre en Algérie." *Forces Armées Françaises* 150 (1959).

Heroys, A. and Léon Thevinin. *L'Armée Rouge et la Guerre Sociale*. Paris, 1931.

Higgins, Major George A. "German and US Operational Art: A Contrast in Manoeuvre." *Military Review* LXI, (1985).

Hiles, Captain David. "Depth and Simultaneous Attack Battle Lab—Fort Sill." *Military Intelligence* (October-December 1993).

Holder, Lieutenant Colonel L. D. "A New Day for Operational Art; May Be Most Important Doctrine Change Since World War II." *Army* 35, no. 3 (March 1985).

Hollis, P. S. "The Army and FA Challenges of Designing Force XXI." *Field Artillery* (December 1994).

Holzer, Robert. "Marine Generals Plot 21st Century Corps." *Defense News* 10 (13-19 February 1995).

———. "U.S. Navy Study Promotes Precision Munitions." *Defense News* 10 (27 February-5 March 1995).

Hooker, Richard D. "Attritionists—or Technologists?—vs Maneuverists." *Parameters* 23 (1993).

———. "The Mythology Surrounding Maneuver Warfare." *Parameters* 1 (Spring, 1993).

Hooton, E. R. *Jane's Naval Weapon Systems*. Issue 15. Coulsdon, U.K. June 1994.

Howard, Michael. *The Franco-Prussian War: The German Invasion of France, 1870-71*. New York, 1962, 1981.

———. *War in European History*. Oxford, 1976.

———. "The Forgotten Dimensions of Strategy." *Foreign Affairs* 57, no. 5 (Summer 1979).

———. "How Much Can Technology Change Warfare?" In Michael Howard and John Guilmartin Jr. *Two Historians in Technology and War*. Carlisle Barracks, PA, July 1994.

Hunt, Barry and Adrian Preston, eds. *War Aims and Strategic Policy in the Great War*. London, 1977.

Isserson, G. S. *Evoliutsiia operativnogo iskusstva*. 2nd edition. Moscow, 1937.

———. *Osnovy oboronitel'noi operatsii*. Moscow, 1938.

———. *Novye formy bor'by*. Moscow, 1940.

———. "Razvitie teorii sovetskogo operativnogo iskusstva v 30-e gody." *Voenno-istoricheskii zhurnal* no. 1 (January 1965).

Ivanov, S. P. and A. I. Evseev. "Voennoe iskusstvo." *Sovetskaia voennaia entsiklopediia* 2 (1976).

Jablonsky, David. *The Operational Art of Warfare Across the Spectrum of Conflict*. Carlisle Barracks, PA, 1987.

———. *The Owl of Minerva Flies at Twilight: Doctrinal Change and Continuity and the Revolution in Military Affairs*. Carlisle Barracks, PA, May 1994.

———. "US Military Doctrine and the Revolution in Military Affairs." *Parameters* 3 (Autumn 1994).

Jaffe, Lorna S. *The Development of the Base Force 1989-1992*. Washington, DC, July 1993.

Jervis, Robert. "The Future of World Politics: Will It Resemble the Past?" *International Security* 16 (1991/92).

Kennedy, Paul, ed. *Grand Strategies in War and Peace*. New Haven, CT, 1991.

Kessel, Eberhard. *Moltke*. Stuttgart, 1957.

Kipp, Jacob W. "Lenin and Clausewitz: The Militarization of Marxism." *Military Affairs*

XLIX, no. 4 (December 1985).

———. "General-Major A. A. Svechin and Modern War: Military History and Military Theory." Introductory essay in Kent Lee, ed., A. A. Svechin, *Strategy*. Minneapolis, 1992.

Kipp, Jacob W., ed. *V. K. Triandafillov's "The Nature of the Operations of Contemporary Armies."* London, 1994.

———. ed. *The Nature of the Operations of Modern Armies*. Translated by William A. Burhans. Newbury Park, Ilford, U.K., 1994.

Knobel, Major Philip E. "Revise FMFM 1, Warfighting." *Marine Corps Gazette* (October 1993).

Korkodinov, P. D. "Kharakter sovremennykh boev." *Voennaia mysl'* no. 2 (February 1941).

Korotchenko, E. "Ob evoliutsii printsipov voennogo iskusstva." *Voennaia mysl'* no. 9 (September 1988).

Krepinevich, Andrew J. "Cavalry to Computer: The Pattern of Military Revolutions." *The National Interest* 37 (Fall 1994).

———. "The Clinton Defense Program: Assessing the Bottom-Up Review." *Strategic Review* XXII (1994).

Kulakov, V.G. "Operativnoe iskusstvo" (Operational art). *Sovetskaia voennaia entsiklopediia* 6 (1978).

Kurochkin, P.A. *Obshchevoiskovaia armiia v nastuplenii.* Moscow, 1966.

LaGare, Major J. Marc. *Paradigm Found—The Nuclear and Nonlinear Battlefields*. Fort Leavenworth, KS, February 1993.

Lamb, Richard. *Montgomery in Europe, 1943-45: Success or Failure*. London, 1983.

Lanir, Zvi. "The Principles of War and Military Thinking." *Journal of Strategic Studies* 16 (1993).

Le Mire, H. *Histoire Militaire de la Guerre d'Algérie*. Paris, 1982.

Libicki, Martin. *The Mesh and the Net: Speculations on Armed Conflict in a Time of Free Silicon.* McNair Paper no. 28. Washington, DC, March 1994.

Lind, William S. "Some Doctrinal Questions for the United States Army." *Military Review* LVII, no. 3 (March 1977).

———. *Maneuver Warfare Handbook*. Boulder, CO, 1985.

Luttwak, Edward N. *Strategy and History: Collected Essays*. Vol.2. New Brunswick, NJ, 1980.

———. "The Operational Level of War." *International Security* 5, no. 3 (Winter 1980/81).

Luvass, Jay. "Thinking at the Operational Level." *Parameters* 16 (1986).

Lykke, Colonel Arthur F., Jr., ed. *Military Strategy: Theory and Application.* Carlisle Barracks, PA, 1981.

Lynn, John A., ed. *Tools of War: Instruments, Ideas, and Institutions of Warfare, 1445-1871*. Urbana, IL, 1990.

Mackenzie, Major General J. J. G. and Brian Holden Reid, eds. *The British Army and the Operational Level of War*. London, 1989.

Madigan, Major James C. and Major George E. Dodge. "Battle Command: A Force XXI Imperative," *Military Review* LXXIV, no. 11 (November 1994).

Marcy, Lieutenant Colonel Scott A. "Operational Art: Getting Started." *Military Review* LXX, no. 9 (September 1990).

Marievsky, I. "Stanovlenie i razvitie operativnogo iskusstva", *Voenno-istoricheskii*

zhurnal, no. 3 (March 1962).
Marill, J. M. "L'Héritage Indochinois: Adaptation de l'Armée Française en Algérie, 1954-1956." *Revue Historique des Armées* 187 (1992).
Martel, A. *Histoire Militaire de la France*. Vol. 4. Paris, 1994.
Maryganov, I. V. *Peredovoi kharakter sovetskoi voennoi nauki*. Moscow, 1953.
May, Ernest R., ed. *Knowing One's Enemies*. Princeton, NJ, 1986.
Mazarr, Michael J. *The Revolution in Military Affairs: A Framework for Defense Planning*. Carlisle Barracks, PA, June 1994.
McAndrew, William J. "Eighth Army at the Gothic Line: Commanders and Plans," and "Eighth Army at the Gothic Line: The Dogfight." *RUSI Journal* 131 (1986), 50-57, 57-62.
———. "Fire or Movement? Canadian Tactical Doctrine, Sicily—1943." *Military Affairs*, 51 (1987).
———. "Operational Art and the Northwest European Theatre of War, 1944." *Canadian Defence Quarterly*. 21 (1991).
McInnes, Colin and G. D. Sheffield. *Warfare in the Twentieth Century: Theory and Practice*. London, 1988.
Metz, Steven and James Kievit. *The Revolution in Military Affairs and Conflict Short of War*. Carlisle Barracks, PA, July 1994.
Moltke, Graf Helmuth Karl Bernhard von. *Gesammelte Schriften und Denkwuerdigkeiten des General Feldmarschals Grafen Helmuth von Moltke*. Dritter Band: Geschichte des deutsch-franzoesischen Krieges von 1870-71. Berlin, 1891.
———. *Moltkes Militaerische Korrespondenz: Aus den Dienstschriften des Krieges 1870/71: Erste Abtheilung: Der Krieg bis zur Schlacht von Sedan*. Moltkes Militaerische Werke, I, Militaerische Korrespondenz, Dritter Theil, Erste Abtheilung. Berlin, 1896.
———. *Moltkes kriegsgeschichtliche Arbeiten: Kritische Aufsaetze zur Geschichte der Feldzuege von 1809, 1859, 1864, 1866 und 1870/71*. Moltkes Militaerische Werke, III, kriegsgeschichtliche Arbeiten, Zweiter Theil. Berlin, 1899.
———. "Aus den Verordnungen fuer die hoeheren Truppenfuehrer (1869)." In *Moltkes Taktisch-strategische Aufsaetze aus den Jahren 1857 bis 1871*. Moltkes Militaerische Werke, II, Die Thaetigkeit als Chef des Generalstabes der Armee im Frieden, Zweiter Theil. Berlin, 1900.
———. "Aufsatz 'Ueber Strategie' vom Jahre 1871." In *Moltkes Taktisch-strategische Aufsaetze aus den Jahren 1857 bis 1871. Moltkes Militaerische Werke*, II, Die Thaetigkeit als Chef des Generalstabes der Armee im Frieden, Zweiter Theil. Berlin, 1900.
Mosher, Major Alan M. *Light Armor In Deep Operational Maneuver: The New Excalibur*. Fort Leavenworth, KS, May 1994.
Mozolev, V. F. "Ob obshchikh osnovakh teorii operativnogo iskusstva." *Voennaia mysl'*. no. 3 (March 1979).
Newcomb, Lieutenant Colonel Curtis L. *Principles of Future Army Force Structure Design*. Carlisle Barracks, PA, April 1993.
Newell, Clayton R. and Michael D. Krause, eds. *On Operational Art*. Washington, DC, 1995.
Nicholson, Jr., Major John W. *Blainey and the Bottom-Up Review: Increased Potential for Miscalculation and War in the 21st Century*. Fort Leavenworth, KS, May 1994.

Niepold, Gerd. *Battle for White Russia: The Destruction of Army Group Centre June 1944*. Translated by Richard Simpkin. London, 1987.

Nikiforov, N. I. "Svechin—Tukhachevskiy," *Gepolitika i bezopasnost'*. no. 2 (1994). "Oboronitel'nyi kharakter sovetskoi voennoi doktriny i podgotovka voisk (sil)." *Voennaia mysl'* no. 1 (January 1988).

Ogarkov, N. V. "Glubokaia operatsiia (boi)." In *Sovetskaia voennaia entsiklopediia, T.2*. Vol. 21. Moscow, 1976.

———. "Voennaia strategiia" (Military strategy), *Sovetskaia voennaia entsiklopediia*. 7 (1979).

———. *Vsegda v gotovnosti k zashchite Otechestva*. Moscow, 1982.

———. *Istoriya uchit bditel'nosti*. Moscow, 1985.

Oliviero, Major C. S. "Manoeuvre Warfare: Smaller Can Be Better." *Canadian Defence Quarterly*. 2 (Autumn 1988).

Paret, Peter. *Clausewitz and the State: The Man, His Theories, and His Times*. Princeton, NJ, 1985.

Paret, Peter, ed. *Makers of Modern Strategy From Machiavelli to the Nuclear Age*. 2nd edition. Princeton, NJ, 1986.

Pavlenko, N. "Nekotorye voprosy razvitiia teorii strategii v 20-kh godakh." *Voenno-istoricheskii zhurnal* no. 5 (May 1966).

Pavlovsky, I. "Sovetskoe operativnoe iskusstvo v Velikoi Otechestvennoi voine." *Voennaia mysl'* no. 3 (March 1968).

"Pentagon Reveals Research Priorities." *Defense News* 10 (27 February-5 March 1995).

Pilsudski, Jozef. *Year 1920*. London, 1972.

Predham, James R. "Army Research Laboratory Contribution to Force XXI." *Army Research, Development and Acquisition* (January-February 1995).

Reid, Brian Holden. "The Italian Campaign, 1943-45: A Reappraisal of Allied Generalship." *Journal of Strategic Studies* 13 (1990).

Reznichenko, V. "Sovetskie vooruzhennye sily v poslevoennyi period." *Kommunist vooruzhennykh sil*, no. 1 (January 1988).

———. "Podgotovka i provedenie armeiskikh operatsii." *Voennaia mysl'*, no. 1 (January 1991).

Richardson, Lieutenant General William R. "Winning on the Extended Battlefield." *Army* 31, no. 6 (June 1981).

Rioux, J. P. *La France de la Quatrième République*. Vol. 1. Paris, 1980.

Rodionov, I. N. "O nekotorykh polozheniakh sovetskoi voennoi doktriny." *Voennaia mysl'*, no. 3 (March 1991).

Roger, General Bernard W. "Follow-on Forces Attack (FOFA): Myths and Realities." *NATO Review* 6 (December 1984).

Ropelewski, Robert. "Army Shifts Modernization Effort from Embedded to Upgraded Approach." *Signal* 49 (November 1994).

Ropp, Theodore. *War in the Modern World*. New York, 1962.

Rosenau, James N. "Armed Force and Armed Forces in a Turbulent World." In James Burk, ed., *The Military in New Times: Adapting Armed Forces to a Turbulent World*. Boulder, CO, 1994.

Rotmistrov, P. "O roli vnezapnosti v sovremennoi voine." *Voennaia mysl'*, no. 2 (February 1955).

———. "O sovremennom sovetskom voennom iskusstve i ego kharakternykh chertakh." *Voennaia mysl'*, no. 2 (February 1958).

Salmanov, G. I. "Sovetskaia voennaia doktrina i nekotorye vzgliady na kharakter voiny v zashchita sotzializma." *Voennaia mysl'*, no. 12 (December 1988).

Savkin, V. Ye. *The Basic Principles of Operational Art and Tactics*. Translated and published by the U.S. Government Printing Office under the auspices of the United States Air Force in the series *Soviet Military Thought*. No. 4. Washington, DC, 1982.

———. *The Basic Principles of Operational Art and Tactics (A Soviet View)*. USAF trans. Moscow 1972, USGPO 1985.

Savushkin, R. *Sbornik materialov po izucheniiu opyta voiny*. Nos. 1—26. Moscow, 1942-1948.

———. *Sbornik voenno-istoricheskikh materialov Velikoi Otechestvennoi voiny, vypusk 1-19*. Moscow, 1949-1968.

———. "K voprosu o vozniknovenii i razvitii operatsii" (Concerning the question of the origin and development of the operation). *Voenno-istoricheskiizhurnal*, no. 5 (May 1979).

———. "K voprosy o zarozhdenii teorii posledovatel'nykh nastupatel'nykh operatsii." *Voennaia mysl'*, no. 5 (May 1983).

———. "Evoliutsiia vzgliadov na oboronu v mezhvoennye gody." *Voennaia mysl'*, no. 1 (January 1987).

Schmitt, M. *De Diên Biên Phu à Koweit City*. Paris, 1992.

Schneider, James J. *The Structure of Strategic Revolution: Total War and the Roots of the Soviet War State*. Novato, CA, 1994.

Schwarz, Benjamin C. *Casualties, Public Opinion and U.S. Military Intervention*. Santa Monica, CA, 1994.

Sendak, Major Theodore T. "The Airborne Antiarmor Defense." *Military Review* LIX, no. 9 (September 1979).

Sergent, P. *Paras-Légion, le 2ème BEP en Indochine*. Paris, 1982.

Sewall, Sarah B. "Peace Operations: A Department of Defense Perspective." *SAIS Review* XV (1995).

Shtemenko, General S. M. *The Soviet General Staff at War 1941-1945*. 2 vols. Translated by Robert Daglish. Moscow, 1981.

Shukman, Harold, ed. *Stalin's Generals*. London, 1993.

Simpkin, Richard. *Red Armour*. London, 1984.

———. "Manoeuvre Theory and the Small Army." *British Army Review* 78 (December 1984).

———. *Race to the Swift*. London, 1986.

———. *Deep Battle: The Brainchild of Marshal Tukhachevskii*. London, 1987.

Sinnreich, Richard Hart. "Tactical Doctrine or Dogma?" *Army* 29, no. 9 (September 1979).

Skachko, P. G. *Slovar' osnovnykh voennykh terminov*. Moscow, 1965. (Translated and published by U.S. Air Force, 1977).

———. "Odnovremennoe vozdeistvie na vsu glubinu operativnogo postroeniia protivnika—vedushchaia tendentsiia v razvitii teorii operativnogo iskusstva." *Voennaia mysl'*, no. 7 (July 1985).

Sokolovsky, S.V. *Voennaia strategiia*. Moscow, 1968.

Solomon, General Leon E. "Shaping the U.S. Army Material Command for Force XXI." *Army Research, Development and Acquisition* (January-February 1995).

Sovetskaia istoricheskaia entsiklopediia. Vol. 6. Moscow, 1965.

Spiller, Roger J. "War History and the History Wars: Establishing the Combat Studies Institute." *The Public Historian* 10, no. 4 (Fall 1988).

Starry, General Donn A. "A Tactical Evolution—FM 100-5." *Military Review* LVIII, no. 8 (August 1978).

———. "Extending the Battlefield." *Military Review* LXI, no. 3 (March 1981).

Sullivan, General Gordon R. "Force XXI: Digitizing the Battlefield." *Army Research, Development and Acquisition* (November-December 1994).

Sullivan, General Gordon R. and Colonel James M. Dubik. *War in the Information Age*. Carlisle Barracks, PA., 1993.

———. "War in the Information Age." *Military Review* LXXIV, no. 4 (April 1994).

Summers, Colonel Harry G. *On Strategy II: A Critical Analysis of the Gulf War*. New York, 1992.

Svechin, A. A. *Strategiia*. 1st edition. Moscow, 1926.

———. *Strategia*. 2nd edition. Moscow, 1927.

———. *Evolutsiia voennogo iskusstva*. 2 vols. Moscow, 1927-1928.

Svechin, A.A., ed. *Strategiia v trudakh voennykh klassikov*. 2 vols. Moscow, 1927.

Sverdlov, F. "K voprosy o manevr v boiu." *Voennyi vestnik*, no. 8 (August 1972).

Swain, Richard M. "The Written History of Operational Art." *Military Review* LXX, no. (1990).

———. *"Lucky War": Third Army in Desert Storm*. Fort Leavenworth, KS, 1995.

Sweetnam, Lieutenant Colonel J. P. "New Thinking in the U.S. Army: The Louisiana Manoeuvres, Battle Laboratories and the Third Wave Army." *Canadian Defence Quarterly* 24 (1994).

Talensky, N. "Razvitie operativnogo iskusstva po opytu poslednikh voin." *Voennaia mysl'*, nos. 6-7 (June-July 1945).

Taw, Jennifer Morrison and Bruce Hoffman. *The Urbanization of Insurgency: The Potential Challenge to U.S. Army Operations*. Santa Monica, CA, 1994.

Taw, Jennifer Morrison and Robert C. Leicht. *The New World Order and Army Doctrine: The Doctrinal Renaissance of Operations Short of War?* Santa Monica, CA, 1992.

Timoshenko, S. K. *Zakliuchitel'naia rech' narodnogo komissara oborony Soiuza SSR Geroia i Marshala Sovetskogo Soiuza S. K. Timoshenko na voennom soveshannii 31 dekabriia 1940 g.* Moscow, 1941.

Todd, Lieutenant Colonel David. "Gird for Information War." *Defense News* 10 (6-12 March 1995).

Toffler, Alvin and Heidi Toffler. *War and Anti-War: Survival in the Dawn of the 21st Century*. Boston, 1993.

Travers, Tim. *How the War Was Won: Command and Technology in the British Army on the Western Front 1917-1918*. London, 1992.

Triandafillov, V. "Vziamodeistvie mezhdu zapadnym i iugozapadnym frontami vo vremya letnego nastupleniya krasnoy armii na Vislu v 1920 g." *Voyna i revolyutsiya*, No. 2 (1925).

Tringuier, R. *Les Maquis d'Indochine*. Paris, 1976.

Trythall, Anthony John. *"Boney" Fuller: Soldier, Strategist, and Writer, 1878-1966*. New Brunswick, NJ, 1977.

Tsvetkov, A. "Operatsiia, ee sushchnosti i znachenie v sovremennoi vooruzhennoi bor'be." *Voennaia mysl'*, No. 3 (March 1955).

Tukhachevsky, M. *Pokhod za Vislu*. Smolensk, 1923.

———. "K voprosu o sovremennoi strategii." In *Voina i voennoe iskusstvo v svete*

istoricheskogo materializma. Moscow, 1927.
———. "Novye voprosy voiny." In *Izbrannye proizvedeniia*. Vol. 2. Moscow, 1964.
Ulianov, V. I. "Razvitie teorii glubokogo nastupatel'nogo boia v predvoennye gody." *Voenno-istoricheskii zhurnal*, no. 3 (March 1988).
"U.S. Military Strategy and Force Posture for the 21st Century." *RAND Research Brief* (September 1994).
van Creveld, Martin. *Supplying War: Logistics From Wallenstein to Patton*. Cambridge, MA., 1977.
———. *Command in War*. Cambridge, MA, 1985.
———. *Technology and War from 2000 B.C. to the Present*. New York, 1989.
———. *The Transformation of War*. New York, 1991.
van Creveld, Martin et al. *Air Power and Maneuver Warfare*. Maxwell Air Force Base, AL, 1994.
Vardamis, Alex A. "German-American Military Fissures." *Foreign Policy* no. 34 (Spring 1979).
Varfolomeev, N. "Strategiia v akademicheskoi postanovke." *Voina i revoliutsiia*, no. 11 (November 1928).
———. *Udarnaia armiia*. Moscow, 1933.
Vasil'ev, V. "Operativnoe iskusstvo kak sostavnaia chast' sovetskogo voennogo iskusstva." *Voennaia mysl'*, no. 6 (June 1965).
Vetoshnikov, L. "Operativnoe iskusstva i ego mesto v sovetskom voennom iskusstve." *Voennaia mysl'*, no. 11 (November 1949).
Vigor, P. H. "Soviet Echeloning." *Miltary Review* 8 (August, 1982).
"Voennoe iskusstvo." *Voennyi entsikopedicheskii slovar'*. Moscow, 1983.
Volkagonov, D. A. and S. A. Tiushkovich. "Voina." *Sovetskaia voennaia entsiklopediia* 2 (1976).
Voloshin, L. I. *Voprosy strategii i operativnogo iskusstva v sovetskykh voennykh trudakh (1917-1940)* Moscow, 1965.
"Teoriia glubokoi operatsii i tendentsii ee razvitiia." *Voennaia mysl'*, no. 8 (August 1978).
Wagner, Colonel Robert E. "Active Defense and All That." *Military Review* LX, no. 8 (August 1980).
Wass de Czege, Huba. "Army Doctrinal Reform." In Asa A. Clark IV, Peter W. Chiarelli, Jeffrey S. McKitrick, and James W. Reed, eds. *The Defense Reform Debate*. Baltimore, 1984.
Wedgwood, Ruth. "The Smart New Weaponry Can Also Help Keep Peace." *International Herald Tribune*, 16 December 1994.
Wiener, Norbert. *Cybernetics: or Control and Communication in the Animal and the Machine*. New York, 1948.
Zakharov, A. N. "Tendentsii razvitiia vooruzhennoi bor'by." *Voennaia mysl'*, nos. 11-12 (November-December 1991).
Zakharov, M. "O teorii glubokoi operatsii." *Voenno-istoricheskii zhurnal*, no. 10 (October 1970).
Zav'ialov, I. "Evoliutsiia c sootnoshenii strategii, operativnogo iskusstva i taktiki." *Voennaia mysl'*, no. 11 (November 1971).
Zervoudakis, A. "L'Emploi de l'Armée de l'Air en Indochine, 1951-1952." *Revue Historique des Armées* 186 (1992).
Zhlobin, V. "Tvorchestvo sovetskogo operativnogo iskusstva v Velikoi Otechestvennoi

voine." *Voennaia mysl'*, no. 5 (May 1950).
Zhlobin, V. and L. Vetoshnikov. "Ob operativnom iskusstve Sovetskoi Armii." *Voennaia mysl'*, no. 3 (March 1947), no. 4 (April 1947).

Index

About the Editors and Contributors

Colonel Charles F. **Brower** IV received his Ph.D. from the University of Pennsylvania and is now Professor and Chair, Department of History, United States Military Academy, West Point, New York. He recently edited *World War II in Europe: The Final Year.*

Sabine Marie **Decup** received her doctorate from the Université Paul Valéry-Montpellier III, Montpellier, France. She has worked as a research assistant (nuclear affairs) at UNESCO Headquarters, Paris, and is now research associate at Université Paul Valéry-Montpellier III. She has published on Anglo-French relations during the Algerian war.

John **English** is the author of the acclaimed *On Infantry* and co-author of *On Infantry: Revised Edition.* He is the main Canadian contributor to *The D-Day Encyclopedia* and his book, *The Canadian Army in the Normandy Campaign*, has just been published in Canadian paperback as *Failure in High Command.* He is currently series advisor to The Praeger Series in War Studies and an adjunct professor of history and war studies at Queen's University and the Royal Military College of Canada.

Professor Michel **Fortmann** has held the chair in Strategic Studies in the Political Science Department at the University of Montreal since 1986. He has published widely in the fields of Canadian and United States defence policy, arms control, intelligence, and decision-making.

Colonel U.S. Army (ret.) David M. **Glantz**, author of *A History of Soviet Airborne Forces* and *The Military Strategy of the Soviet Union: A History* is the former Director of the Foreign Military Studies Office, Combined Arms Command, Fort Leavenworth, Kansas. He has published widely on Soviet military theory and practice and is the founder and U.S. editor of *The Journal of Slavic Military Studies.*

Thierry **Gongora** has his doctorate from Carleton University. He is currently a research associate with the Chair of Military and Strategic Studies, University of Montreal. He publishes mainly in the areas of modern strategy, international relations, and Middle Eastern Affairs.

Michael A. **Hennessy** is an assistant professor of history and war studies at the Royal Military College of Canada, where he specializes in naval history, military technology and low-intensity conflict studies. He is the author of the forthcoming Praeger title, *Strategy in Vietnam: Revolutionary War in I Corps, 1965-1971.*

Jacob **Kipp** is the Senior Analyst and Research Coordinator, Geo-Strategic Studies Office, Foreign Military Studies Office, Fort Leavenworth, Kansas, and adjunct professor of history, Kansas State University. The author of a large number of articles and monographs on Russian and early Soviet military history, he is also the U.S. co-editor of the quarterly journal, *European Security.*

Stéphane **Lefebvre** completed doctoral studies at the University of Montreal and has worked as a strategic analyst for the Canadian Department of National Defence. He has published widely on problems of Balkan security.

William (Bill) **McAndrew** is recently retired from the Directorate of History, Canadian Department of National Defence, and was formerly the academic director of the Canadian Forces Staff College. He has published widely on Canadian military history, particulary operations during the Second World War. With Terry Copp he is author of the acclaimed *Battle Exhaustion: Soldiers and Psychiatrists in the Canadian Army, 1939-45.*

B.J.C. **McKercher**, Professor of History at the Royal Military College of Canada, has written widely on Anglo-American and Anglo-Canadian relations in the twentieth century. He is general editor of Praeger's Series on Diplomacy and Strategic Thought, and recently co-edited (with Lawrence Aronsen) *The North Atlantic Triangle in a Changing World. Anglo-American-Canadian Relations, 1902-1956.*

Bradley J. **Meyer** received his Ph.D. from Ohio State University, taught there as a sessional lecturer, and is now Associate Professor of Military History at the United States Marine Corps' School of Advanced Warfighting, Command and Staff College, Quantico, Virginia. He has published several articles on Moltke the Elder, von Schlieffen, and the operational art. He is now completing a book on the Elder Moltke's influence on Prusso-German operational art.

Colonel U.S. Army (ret.) Richard M. **Swain**, late of Fort Leavenworth, is the Director of the Combat Studies Institute, U.S. Army Command and Staff College, Carlisle Barracks, Pennsylvania. Col. Swain has written widely on the development of military doctrine and served as the U.S. Army's Theatre historian for Operation Desert Storm.

ISBN 0-275-95305-X

90000>

EAN

9 780275 953058

HARDCOVER BAR CODE